REBOOT
NATION

A Guide to the Internet for the Technically Challenged

REBOOT NATION

A Guide to the Internet for the Technically Challenged

Steven Grabiel

ARPress

ILLUMINATING IDEAS
EMPOWERING VOICES

ARPress
45 Dan Road Suite 36
Canton MA 02021

Hotline: 1(888) 821-0229
Fax: 1(508) 545-7580

Ordering Information:
Quantity Sales. Special discounts are available on quantity purchases by corporations, associations, and others. For details, contact the publisher at the address above.

Printed in the United States of America.

ISBN-13   Paperback   979-8-89330-813-6
             eBook   979-8-89330-814-3

Library of Congress Control Number: 2024902801

This book is for God's Glory Alone

REBOOT NATION
BY STEVEN GRABIEL

Everyone needs internet access. Understanding the different types available and exactly what an internet service provider is offering is the purpose of this brief book. Most chapters are dedicated to a particular kind of internet service, and each gives a basic explanation of the logistics of the offering and its pros and cons. The last two chapters provide the reader with helpful tips for getting the most out of their internet and ways to find internet services available in their area. Following this are a few appendices that go into more depth on more technical areas.

Grabiel has many years of experience in the field he is writing about. He uses this experience and the lessons he has learned from dealing with customers over the years to guide him while writing a beginner-friendly guide to understanding and purchasing internet service. The writing is clear and to the point, and the book is separated into easily manageable chapters. Grabiel uses plain speech and offers examples to help one understand the concepts he explains. At times, the presentation is a bit dry, and there are instances when a one-page infographic might display the same information in an easier-to-grasp and remember format. Still, sitting well under one hundred pages, the book delivers the information in a concise format that is understandable to most readers. Anyone wishing to understand the world of internet service providers to be better informed and make an intelligent decision before making a purchase will find good information in this book.

-US REVIEW OF BOOKS

CONTENTS

ABOUT STEVEN GRABIEL

Steven's clients call him the Internet Guy.

Steven began poking around with computers and the infant Internet in 1982. Working a summer job doing landscaping, he saved up enough money to purchase an Atari 400 computer and a modem. Computers back then plugged into a television for the monitor, so he plugged this computer into his black-and-white set. He then plugged in a modem into the telephone jack and started his journey into the Internet by subscribing to CompuServe. Back then, there were no such things as graphics online, just text. The Internet was a novelty; there was not even email yet. All a person could do was communicate with other users around the globe in a chat-like environment, similar to what we know today. That was dial-up Internet.

Steven attended California State University in Long Beach beginning in 1988 in pursuit of a degree in electrical engineering. After three years he transferred to California State University in Chico to finish up his degree. College was as struggle, but he eventually graduated and continued to gain knowledge about the growing Internet. While working toward the electrical engineering degree, he also took an interest in microwave transmissions of signals. This interest would blossom into the business he now operates.

In 1995, Steven went to work in his father's electrical contracting business in Albuquerque, New Mexico. He started as a "shop boy," a position he had also held in 1986.

His first task for the business was to modernize their computer systems and create a network for the computers to share resources.

After that was complete, he began to modernize the business's billing systems from a paper method to a database method. This single task vastly improved cash flow, and growth for the business was incredible. At that time, the business was not connected to the Internet, as it was still a novelty to Steven's father and his partners.

But in 1996, Steven ushered the Internet into the business by tapping into the fax dial tone and setting up a flash.net account. The fax machine was heavily used at that time because faxing information was the standard of sending documents. Steven was often told to get offline so someone could send a fax. Little did they know the fax machine would soon become a dinosaur.

In 1999, Steven brought broadband Internet to the business. No longer did the fax line need to be tied up to communicate via email, plus it was faster than the original dial-up service. The connection he brought in was a fixed-base wireless Internet connection. The speeds were 128 k or double the speed of the fastest dial-up. This was the most economical form of broadband because a dedicated T-1 cost a thousand dollars a month. Although Steven's father never adopted the computer or Internet, his partners soon saw that the Internet was not just a novelty and a pest that tied up the fax line but a way in which modern business was heading.

Steven left the contracting business in 2002. With his knowledge in computers, networking, and what broadband Internet could do, he believed there had to be a way to provide broadband for everyone, just like the Internet he obtained for his father's electrical contracting business. In 2005, Steven started Internet Services LLC, doing business as Higher-speed Internet. The business was developed to provide fixed-base wireless Internet in the area he lives in (more on fixed-based wireless Internet in chapter 5). Since 2005, Internet Services LLC continues to provide broadband Internet to the East Mountains of New Mexico.

In addition, Steven hosts an Internet-based radio show that meets weekly to discuss things of interest to Internet service providers. The show can be heard live via podcast at www.ispradio.com.

INTRODUCTION:
HOW TO USE THIS GUIDE

Reboot Nation: A Guide to the Internet for the Technically Challenged explains the current technologies that deliver fast Internet to you in simple-to-understand terms. This guide will offer recommendations on what is considered the best broadband Internet connection for the price, if multiple choices of providers are available in your area. This guide will explain what costs you can expect to pay for a given broadband connection. This guide will also explain the technical difficulties that may arise with the use of the different forms of broadband Internet that cause frustration. The overall buying process would not be complete without a discussion on customer service. The pros and cons of each technology will be examined and explained.

Some of the people I serve tell me that they are "technologically illiterate." They are smart people, but they are not interested in the geek-speak technological terminology. If that describes you, then this book is for you. My goal is to empower the self-proclaimed "technologically illiterate" reader with a basic understanding of the technologies that provide fast Internet, what they could expect to pay for each service, who typically provides the services, and who typically will buy the services. Some locations have many fast Internet options available, and some do not. After reading this guide, you will be able to discern the differences between the various ways fast Internet is delivered to your location.

Because I meet all the people I bring fast Internet to, I have learned that a majority of folks just want a reliable, fast, and inexpensive Internet connection. I explain daily to my customers easy-to-understand terms and analogies to common Internet jargon. For example, one effective analogy to understand broadband Internet is to think about your water system. Just like a water company, in the Internet industry we build pipes. And just like the water company, if there are large pipes, you receive plentiful water (quantity) at a good pressure (quality). The same is true in the Internet industry. The size of the pipes we build determines the speed by which you receive the information, both quality and quantity.

Techno-savvy salespeople love to boggle the minds of prospective buyers. Overwhelmed with techno jargon, the prospective buyer just nods their head. T-1, OC3—what does that all mean? Is that really important in getting a fast Internet connection? I can proclaim, "No! It is not necessary!"

Another concept that must be understood from the beginning is that when you purchase any form of Internet connection, you are paying for two services. Many companies, from the telephone company to the cable company to the WISP, bundle these two together. They may bill you separately for the two services on one bill, and that is why the bills from the telephone company are so complicated.

The two services you pay for when you purchase an Internet connection are the following: transport and Internet connection. Transport is the physical medium that the data rides along (e.g., wire, fiber optics, air, etc.) to get from point A to point B. With a dial-up connection, the transport portion of the payment comes with your telephone bill, which you experience as a "dial tone." The Internet portion of this service is paid to AOL, for example. AOL sends you a monthly bill for Internet access because you dial an AOL telephone number and get online through that telephone number. If you have DSL with your local incumbent telephone company, then you are paying for the DSL signal separately from the Internet access. Take a look at the line items on the phone bill. You can pay for another Internet provider in the same way as in the dial-up AOL example, and possibly save some money. However, bundles from providers do not necessarily save anyone any money in the long run.

I shall begin with some common terms explained in easy-to-understand ways, often using such analogies. The terms given are not an exhaustive list, just a minimal list of terms that you as a consumer of broadband Internet should know. These terms are necessary to understand so that you can keep your provider in check to ensure you are receiving what they promise, as well as be able to speak and understand some of their terms.

This guide will then explain the different technologies out there that deliver Internet, from the earliest forms of Internet connectivity to the more modern ways we connect to the Internet.

But what good would this book be if you have all this knowledge and not know what is available to you in your location? Chapter 109 explains how to use some tools to find the most cost effective and overall best solution for you.

There is money to be saved by the technologically challenged in this industry, and you will be able to do just that after reading this book.

Why Reboot Nation? Rebooting of the equipment you use to be online e.g., your computer and your modem, is the first thing you should do if you cannot connect online. Rebooting simply means shutting down your computer or unplugging power to your modem. Every help center technician you call will ask you if you have done this. The reason why is rebooting refreshes the computer and modem and clears up a majority of trouble you may have.

CHAPTER 1
Non-Geek-Speak Terms

The following list includes terms you most need to know, because they will be used regularly in this guide, and you may need to use them to communicate with Internet service providers (ISPs). As previously mentioned, this is not an exhaustive definition list.

Broadband Internet = Fast Internet

Bandwidth. The rate of data transfer, bit rate or throughput, measured in bits per second (bit/s).

Bits. All Internet, accessed by no matter what technology, is simply the turning on and off of an electrical circuit, much like turning on a light and turning it off. A bit is turning that light on and off once.

Kilobits
Turning that light on and off a thousand times.

Megabits
Turning that light on and off one million times.

Gigabits
Turning that light on and off a billion times.

Bits per Second (bps). Turning that light on and off once in a second.

Kilobits per Second (Kbps)
Turning that light on and off a thousand times in a second.

Megabits per Second (Mbps)
Turning that light on and off a million times in a second.

Gigabits (Gbps)
Turning that light on and off a billion times in a second.

1

Byte. A group of bits packaged together by the computer to be transported from one computer to another.

Demarcation Point. The point where your ISP is no longer responsible for your service. (For more information, see appendix C.)

Digital. The on-off switching.

DSLAM. Digital subscriber line access multiplexers

Firewall. A firewall is a network security system that monitors and controls the incoming and outgoing network traffic. No firewall is perfect in keeping out things that you don't want to be there (such as malware).

Head End. A run of cable from a cable company's head end to your home and into the "cable box" that sits on your TV.

Incumbent Telephone Company. The old telephone company we all know: Pacific Bell, Western Electric for Mountain Bell, Southwestern Bell, AT&T, Qwest, CenturyLink, etc.

Internet. Just a connection between computers.

ISP. Internet service provider.

WISP. Wireless Internet service provider.

Latency. The time it takes for a byte of information to get from one computer to another.

LAN. Local area network.

WAN. Wide area network.

Malware. Stands for malicious software. It is any software used to disrupt computer operations, gather sensitive information, gain access to private information on computers, or display unwanted advertising.

Medium. How the Internet is delivered to your location (through copper telephone wires, for example).

Modem. A computer (without a keyboard or mouse) that converts the "on-off light switch" to something that is useable to your computer.

Packet. A group of bytes packaged together for transmission through a network.

PoE. Power over Ethernet. Any device that brings electricity and data to your Ethernet cable.

POTS Service. Plain old telephone service. This kind of service does not have the voicemail, caller ID, call waiting, long distance, etc. It is as the name implies: basic telephone service.

Reboot. Turning your electronic devices and equipment off for a short period (minutes) and then on again. This allows the device to reload the operating system. It can clear out things you don't want STEVEN GRABIEL in there and help improve connections and speeds. (See tips chapter for more information.)

Router. Similar to a modem because it is what helps to make a useable connection from your computer to the Internet.

Server. A server is a computer program that provides services to other computer programs. Most servers are housed on dedicated computers with a lot of bandwidth so they can serve many other computers and their users (like you) from all over the world.

Speed Test. This is a test ISPs use to determine whether they are providing you what you pay us for. We use www.speedtest.net. You can use this too. Results will be given as Mbps, or megabits per second.

T-1. A copper-based dedicated Internet pipe that is not shared with anyone, so many corporations may still have them.

Wi-Fi. Wi-Fi stand for Wireless Fidelity, it is a local area network technology that allows electronic devices (computers, cellphones) to connect to a network without wires.

CHAPTER 2
Internet Prior to 1995, Dial-up, and T-1 Internet Service

"I have been on dial-up for so long."

"Really? I didn't think people signed up for dial-up Internet anymore."

"Yeah. It's slow, but we didn't know of any other alternatives. I log in and go to my bank's website, and while it is loading the site, I go make a sandwich and have lunch. By the time I am done, the site is loaded. I just use the Internet for email but even that has gotten so slow. I don't have the time to be online for this."

"I understand. Let's see if we can get you a faster connection that is within your budget."

This is a conversation I have with many people. So, what is going on with dial-up Internet?

Telephone wire

Medium: One pair of copper wire.

Dial-up pros: Anywhere there is a dial tone, there can be Internet. You only need a dial-up modem and an account with a provider (landline telephone company) to dial into.

T-1 pro: Similarly, anywhere there is dial tone, there can be this form of Internet.

Cons: Dial-up is a slower connection to the Internet compared to the alternatives. T-1 Internet is faster but can be expensive and has long-term commitments for better pricing. Additionally, the age of copper wire infrastructure in this country can create frustrating problems.

Cost: Dial-up typically runs from $9.95 to $24 per month depending on the Internet provider and cost of dial tone from the telephone company. T-1 will cost from $300 to $900 per month for the telephone company connection plus fees from the Internet provider.

Speed expectations: Dial-up is 26 Kbps to 56 Kbps. T-1 1,536 Kbps.

Many years ago, prior to any Internet, there were party lines. We called them party lines because you could pick up the phone and hear a conversation going on that you were not a party to. We had party lines because a pair of wires went into every home. That wire, on the other end, would go to a central location in your neighborhood; then it would go to the outside world. At that central neighborhood location, all the wires were tied together, so when you picked up the phone, if no one happened to be on the line, you could make a call. Sometimes you would pick up the telephone and your neighbor was on the line speaking with someone. The industry (the Bell Company) implemented something called a T-1. (Who knows why it was called T-1. It is geek speak.)

T-1 is a circuit that has 24 telephone lines in it. It is not a bundle of telephone wires put together. They use a technology called multiplexing, which is able to put 24 telephone lines onto one telephone line. Party lines ended with the implementation of this technology. When you picked up the telephone, you were able to get a dial tone almost every time. In the infancy of the Internet (post party line), a standard one-pair copper telephone line at best was able to produce 64 Kbps of data transmission (analogous to turning on and off the light switch 64,000 times in one second). This holds true today. The reason I say at best is there are many factors that made this slower at your home, such as the size of the copper wire to your home and the distance from the neighborhood central location. It just so happened that in the days prior to the Internet, the dial tone you used to talk to your neighbor used a technology called analog that happened to also switch at 64 Kbps.

To speak with someone, you needed a telephone plugged into the telephone system. To communicate with someone using a computer back in the beginning of the Internet, you needed a dial-up modem.

Dial-up Internet is still in existence, but with the more modern technology being able to switch faster, from at best 64 Kbps to millions of switching per second, a dial-up connect can be a good backup solution should you live and die on the Internet.

The cost of dial-up Internet is extremely inexpensive and can possibly be free, depending on the provider. The technology may not be sustainable in the future, as other forms of data transmission are becoming more prevalent and more cost effective. It isn't there yet, but this technology might very well go the way of the dinosaur.

What you need: To receive dial-up connections to the Internet, you must have a pair of copper wires from your home to the telephone company and pay the telephone company for the telephone service. You also have to pay an Internet provider who will provide you with a dial-up telephone number to connect to the Internet. The equipment needed for a dial-up connection is a simple modem that can be internal or external to your computer, depending on the model, and a telephone cord to plug the modem into the wall for dial tone. Sometimes the telephone company can be your Internet provider, but that is not a majority of the cases (e.g., AOL, EarthLink, etc.).

The biggest problem with dial-up that can cause frustration is its slow speed. It was great as a novelty back in the infancy of the Internet, but now, time is valuable.

Also, the equipment used and the infrastructure (copper wire) that delivers the service is very old. Remember that Alexander Graham Bell first created a working telephone in 1876, and much of the copper wire infrastructure in the United States is now at least 100 years old.

That said, dial-up is ideal as an inexpensive backup connection to the Internet, should the other forms of Internet you have on occasion fail.

The T-1 service mentioned earlier was, and still is, what some industries use to connect to the Internet. A T-1 is a dedicated single pair of copper telephone wire that offers 24 telephone lines through multiplexing them all together onto one copper wire. With 24 lines over one wire, the total switching per second increases from 64 Kbps for a single line to 24 lines times 64, which equals 1,560 Kbps. A faster rate of switching speed means faster Internet.

The delivery of a T-1 is the same copper wire that is used to connect directly from your location to the telephone company's main building. Remember, the dial-up connection is from your home to the neighborhood communications box.

There are two devices used for this technology to provide an Internet connection. One device is called a network interface unit (NIU), and this is owned by the telephone company. The other device is called a router, which you own (or the telephone company may lease to you if you want). The router is really the modem because it is what helps to make a useable Internet connection from your computer to the Internet.

A T-1 is what the industry calls a leased line. This means a long-term contract with a hefty price per month. What you are paying for with this service is the copper wire's distance from your home to the telephone company's main office. These prices are all tariffed (regulated by the Federal Communications Commission and your state).

Similar to the dial-up connection, the actual telephone connection is only the first half of the equation when connecting to the Internet. You still need that connection to an Internet provider. Similar to the dial-up connection, the telephone company can be that provider, but that is not necessarily the rule.

Today the price for a T-1 can run from $300 to $1,800 per month, depending on the distance from your location to the telephone company's main office.

Both dial-up service and T-1 service, the oldest form of Internet connection, are still with us. What are you paying for when you look at a T-1 for your Internet connection? As mentioned, you are paying for a direct connection from your location to the telephone company's main office and the mileage. Also, T-1 service being regulated by the government, usually has a long-term contract in place. With a long-term contract you receive a service level agreement (SLA). This is a document where the provider will guarantee speeds and offer guarantees for uptime. The price point of the T-1 is probably out of the range of all of us non-multimillionaires, but it is an alternative for those who absolutely need a fast Internet connection and live outside the reach of the other forms of Internet connection to be discussed.

T-1 service is not perfect. It is a technology developed by humans, so it can break. Frustrations induced by no connection can occur. Lightning has been known to bring this service to its knees (other services as well). You are paying a hefty fee for this service, but because you have an SLA in place, if for any reason the service is interrupted, you have the comfort of knowing you can be compensated for downtime.

Installation of a dial-up connection is usually just plugging in the modem onboard the computer to the wall jack where there is dial tone. Sometimes special programming is necessary to get online, but most of your large dial-up providers provide a CD you put in your computer that walk you through the process.

Installation of a T-1 service requires a special router with a device called a CSU/DSU (Channel Service Unit/data Service Unit. Programming this router is not for the layperson. Because of the cost of a T-1, most T-1 providers will lease you a router or you can provide one. Sometimes the provider of the T-1 service will program the router for access to the Internet if the router is leased. If you chose to purchase your own router, you will need someone savvy in programming the router for connection to the Internet.

Customer service for either dial-up or T-1 can be frustrating in itself because you are usually dealing with someone not geographically located near you. The providers of these services are usually the large telephone companies and large dial-up Internet providers. If you are able to talk with an agent to help you, they may be in a foreign country or speak in a way that you cannot understand.

The conversation I shared at the beginning of this chapter ended up with a great outcome. The lady lived in the middle of nowhere, but in a place where there was plain old telephone service. She needed a faster Internet connection because time was of value to her. When we got her the faster Internet she wanted, we found her computer was so old that she had to upgrade that too. So, keep in mind the newest Internet technologies work with the newest computers. Take care of that too, and you will have the faster service you want.

CHAPTER 3
Digital Subscriber Line (DSL) Internet Service

Medium: One pair of copper telephone wire.

Pros: Reliable broadband Internet connection typically.

Cons: Distance from electronics head end makes for slower-than-advertised speeds. Aged copper telephone wires can reduce speeds. The cost for a dialup connection is $19.95 to $24 per month depending on Internet provider. Usually imbedded in a complex monthly bill.

Speed expectations: 1.3 to 40 Mbps depending on distance from electronics, age of infrastructure, and technology deployed in the field.

A few years following the growth of the dial-up connection, the development and adoption of such applications as email began to bog down connections to the Internet. Computers became more affordable, and more people were getting computers at their home. People became impatient with their time spent online because of the slow rate of connection that dial-up Internet provides. The novelty of being able to communicate with someone via email was becoming a utility. People who worked for large corporations that had expensive T-1 connections to the Internet experienced what fast Internet was all about. Faster connection in the home was needed.

The telephone companies began to roll out a product called DSL, digital subscriber line. This technology allowed a faster connection to your location over the same copper wire you were using for the dial-up connection. Telephone companies were able to put electronics at the neighborhood connections boxes you may see in the field that puts the data on the same line that provides your dial tone. The electronics they put in the field in those boxes are called DSLAMs (digital subscriber line access multiplexers). This was amazing at the time they did this. There were two services over the same copper wire: your voice service and your connection to the Internet.

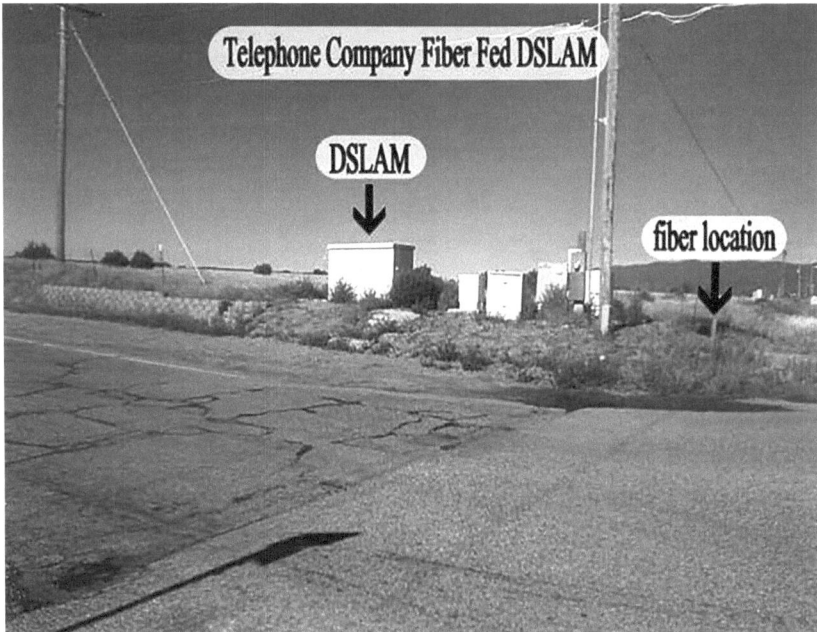

Speeds increased from the best dial-up connection of 64 Kbps to the T-1 speeds of 1,560 Kbps. For the homeowner, this was a huge leap.

You may think that this leap would come with a large price tag, but DSL service was not as expensive as the T-1 service. How could this be? Remember, the majority cost of a T-1 is the distance from your location to the main telephone office. That distance is not measured as the crow flies. It is the path the wire takes to get to your location. With DSL service, the distance does not impact the price. The reason is that the electronics they placed at the neighborhood connection boxes do not require a direct connection from your location to the telephone company's main office, just from your location to those electronics, the DSLAM.

What then, you may ask, is the downside to receiving T-1 service at a much lower cost? The distance of your location to the DSLAM affects the performance. The further away from the DSLAM your location is, the weaker the signal will be. The weaker the signal, the less likely the telephone company will be able to deliver the speeds they advertise. If you are close to the DSLAM, your Internet performance may be near T-1 speeds. In some circumstances, your location being physically closer to the DSLAM does not necessarily mean the best connection. The reason I say this is that wire might not go directly to the DSLAM. The wires to your location may take a roundabout way of getting to the DSLAM. This is a rare occurrence, but I have seen it in the field.

Another reason for poor-performing DSL service is the size of the wire from the DSLAM to your location. The telephone company installed those wires to provide your telephone service many years ago. There was no Internet to think about. The wires they used were completely adequate to provide telephone service but may not be adequate to provide Internet.

DSLAM equipment are sometimes provided their upstream connectivity to the internet via copper based T-1 connections (usually several) and some are provided this upstream connection via fiber optic connection (more on that in chapter 8). What does this mean? If the DSLAM in your community is fed from one T-1 data circuit, the DSLAM has the ability to provide 1,560 Kbps. If you are online and you are the only one online in your community and you fire up a download, you will receive all the speed that that T-1 that feeds the DSLAM is capable of providing. If your neighbor happens to be online at the same time and doing the same thing you are doing; speeds will be decreased.

There is a water analogy to this. Say you are in the shower, and you have good pressure and good temperature. Someone starts up a load of laundry using hot water, and at the same time someone starts to do the dishes. The pressure and volume of hot water will be reduced to your shower as well as at each place someone is trying to use the hot water, because the same amount of incoming water is now going to three different places. The same can be said for data going across an Internet connection. If you are downloading your favorite TV show and your child is watching a video online, the person who tries to access a bank website may experience a slow response.

The electronics you need for DSL service is also a modem or a router. That modem converts the DSL signal to useable Internet information for your computer. A router can be purchased inexpensively ($50–$100). The telephone company may also lease a modem/router to the end user. Setting up a DSL connection is usually quite simple. The routers will come with an installation CD that has a "wizard" that will take you, step by step, through the process.

Your DSL provider will provide you with credentials that will need to be programmed into the router so that it can communicate with the Internet. Credentials will include a username and password that the manufacturer assigned to that particular piece of equipment. I highly suggest you keep those credentials handy should the modem experience a major failure where you need to replace it. Without those credentials you may be on the telephone with your provider for a long time trying to get online. Sometimes the DSL provider will dispatch a technician to get the service up and running at your location for a small fee. Different telephone companies handle new DSL installations differently.

Today, connection speeds to the Internet delivered by the telephone company can be from T-1 speeds up to 40 Mbps download speeds. It all depends on what equipment the telephone company deploys at the DSLAM, such as the latest technology within the DSLAM, and whether the upstream bandwidth is provided by a fiber-optic connection or multiple T-1 copper connections.

Cost of this service from the telephone companies can be deceptive. If you bundle multiple services together (voice and Internet), they sometimes claim that the DSL is a fixed low price for many years. What they neglect to mention is the voice component of their services is what gets hammered with taxes etc. and that the voice service is the service they are constantly going back to regulators requesting rate increases. If you can get by without the need for the dial tone component of a DSL circuit, you can expect to pay $40 to $50 per month for this kind of broadband Internet. If you have to bundle voice and Internet together and want to save the most money, request that the dial tone be just a POTS (plain old telephone service). This kind of service does not have the voicemail, caller ID, call waiting, long distance, etc. It is as the name implies: basic telephone service. A bundled telephone and Internet service via DSL can be as low at $55 per month (without the taxes) if the voice component is a POTS line.

Like the dial-up and T-1 service, age of infrastructure, the size and location of the copper wire, the DSLAMs, and what the DSLAM's upstream service is will determine how reliable the service is. If the copper is old and smaller than the newly deployed copper wire or you are at a distance of 12,000 feet or more as the wire goes to your home, you will experience slower-than-advertised speeds to the Internet. It is the physics of copper wire and Ohm's Law. If that DSLAM is fed by one T-1 upstream connection, you will experience slowness due to congestion. If the connection box is old and has a lot of dust in it, and moisture makes its way into the box, you will experience disconnections to the Internet.

In some areas you can verify whether there is DSL by asking your neighbors. The telephone company might tell you there is no such service. This is because the telephone company may have only deployed a DSLAM with so many ports and they are all taken up by paying people. When someone moves or stops service, that frees up the ports available for additional growth. If demand for a specific area dictates more ports, the provider will usually add equipment to handle the growth, but remember you are usually dealing with a behemoth of a company that is never fast to do anything.

Customer service is usually handled by someone reading off a script that does not reside in the same state as you. If your problem does not coincide with what their script says, you will probably be transferred up the chain of command. Dealing with large companies, you can expect their never-ending calling queues. (Press 1 for this and 2 for that. Please wait to be transferred to the next available agent in three days.) To me, that is the most frustrating aspect of dealing with DSL providers. You just want to talk with someone who speaks the same language as you and knows how to handle your situation.

CHAPTER 4
Cable Internet Service

Medium: Coaxial cable.

Pros: Reliable broadband Internet connection typically.

Cons: Not everywhere and can be expensive where available.

Cost: Internet alone can cost from $50 to $65 per month depending on Internet provider. Usually imbedded in a complex monthly bill that is bundled with other services such as voice and television. When adding their additional services, the monthly bills can exceed $200 per month.

Speed expectations: 5 to 50 Mbps download depending on what the cable provider is willing to sell you.

In the late 1970s and early 1980s, smart folks learned how to transmit television programming across a medium called coaxial cable. The technology required what they call a head end, a run of cable from their head end to your home into the "cable box" that sat on your TV. Because it was not broadcast out into the open air like free TV, they could broadcast some things not appropriate for all audiences, unlike free open-air TV. Back in those days, small independent cable companies started to string the coaxial cables to neighborhoods in search of customers, running their coaxial cable below ground in conduits or above ground on power poles from their main offices to your home. In the mid-nineties, these people who had this infrastructure in the field figured out the medium they were running this on, the coaxial cable, was ideal to transport fast broadband Internet (also their television service), just as the telephone wire was able to transmit not only voice but also DSL broadband.

Today there seems to be fewer independent cable companies. Many have been swallowed up by larger cable companies. The technology they used to deploy the television over the coaxial cable has changed a little. What the cable companies are doing now that was different from when they started is they are replacing some of the coaxial cables running from their head end to neighborhood consolidation points with fiber optics cable, much like the telephone companies' DSLAMs.

From these areas they convert to their existing coaxial cable. The technology model is similar in regard to how the cable companies promote their service packages e.g., multiple services over one cable, just like the telephone companies promote theirs. The difference is in the last-mile connections from their community boxes to the home.

The cost of a cable Internet connection should run $50 to $65 per month. As with the telephone company, they bundle their services together to make you feel you are getting a great deal. I have seen cable company invoices that included Internet, television, and telephone services, exceeding $200 per month. If you are on a smaller budget and do not want the other services, just getting the Internet can be feasible without the other services they will try to sell you.

The equipment needed in your home is a modem. It converts the coaxial cable service to Ethernet service. Usually, it is just a one modem/one connection device with no Wi-Fi (or wireless capabilities) built in. The cable companies are now getting better and are selling to the end user equipment that provides Wi-Fi capability. The cost of these cable modems can be from $50 to $100 depending on the capability you desire (such as a Wi-Fi connection or more than one Ethernet port).

Setting up a cable company connection is usually easy. The cable modems will come with an installation CD, like the DSL providers provide with their routers. Like the DSL routers, your cable company may send you credentials you will need to program into the modem for the connection to work.

Cable modem with built-in Wi-Fi

Be sure to keep the credential information for future reference should something catastrophic occur with your cable modem. Sometimes the cable company may send a technician to get your service set up for a small fee.

The speeds you can expect to receive from the cable companies differ in different marketplaces. The following picture is of a speed test through a cable company's connection.

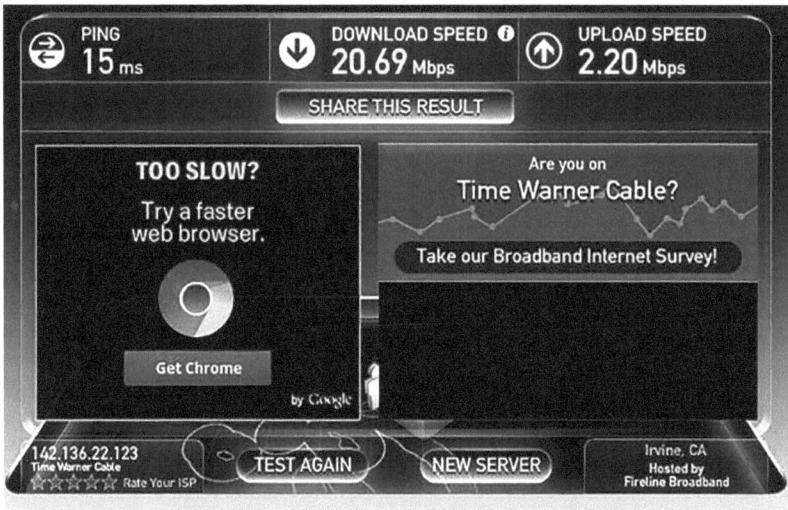

This is a residential Internet connection in Southern California that I used to pay about $50 per month for, and this is just their Internet connection. Great for the price!

Problems faced with cable company connections are generally limited to direct lightning strikes to their equipment in the field or their cables running on power poles. Debris from things falling and hitting the cable on the poles and destroying the cable (such as broken tree branches in a storm) can cause problems too. By and large, the connection is relatively new compared to the telephone company's connection, so age of infrastructure is not really an issue. The cable company has created a decent service that has grown from just providing television service.

The customer service of these providers is usually from someone not in your area and not aware of the local circumstances you may have, because most of the cable companies you meet are large multistate companies, much like the telephone companies. Customer service from these large companies usually causes the most frustrations, similar to the telephone companies.

CHAPTER 5
Fixed-Base Wireless Internet Service

Wireless network

Medium: Air.

Pros: Speeds can typically be greater than most of the providers previously mentioned already, depending on what you want to pay. A nice alternative to the big company band width and usually provided by a small business whose main objective is to provide you the best service they can.

Cons: Weather and terrain obstacles can impair service.

Cost: Dial-up $25 to $100 per month depending on Internet provider and what speed you want.

Speed expectations: 1 to 100 Mbps download depending on what the provider is willing and capable of selling you.

In the early to mid-'90s, this new form of fast Internet was coming on the stage, called fixed-base wireless microwave Internet. What does fixed-base microwave mean? This form of Internet is typically broadcast from a central location to a tower and then rebroadcast and distributed to end users by way of a radio attached to their location. Instead of using wires like the telephone or cable companies use, the medium of transmission is air. This is similar to how you receive free over-the-air radio and television. You have a television connected to an antenna that receives the radio transmission. The antenna is fixed to your home.

The difference between free over-the-air radio or television service and fixed-base wireless Internet is that the device on your home talks back to the other end. Your television and radio do not talk back to the television or radio network.

Fixed-base wireless broadband Internet is the industry that I belong to. The cost and time to deploy an Internet network such as this is substantially less than anything that requires expensive wire or cables strung from the Internet providers' location to your home. Transmission of information is all taken care of for free through the air! This is the reason I became involved in the industry (and why it will be the longest chapter in the guide). I can deliver a quality broadband connection to rural parts of New Mexico that the telephone and cable companies find no economic interest in delivering services to. The capital cost to bring their services to rural areas is extremely high, and if the population density of a rural area does not meet their criteria, they will not invest in the area. It is all about economics. If four people in a remote area want broadband Internet but the cost of bringing a wire or cable to the area is thousands of dollars, the telephone and cable companies cannot justify the cost to their investors. Recoupment of the investment takes too long.

The classification that folks who deliver this form of broadband Internet are commonly called WISPs, or wireless internet service providers. Similar to the small independent cable companies who invested years ago in delivering television service to the home by buying cable and installing it on power poles or in underground conduits, WISPs are typically small independent shops run by entrepreneurs who do not have vast sums of capital to deploy expensive wires or cables. Most WISPs started with a T-1 connection for upstream connectivity and converted that form of Internet to microwave Internet for distribution. As WISPs grow, they usually add additional T-1s, then possibly invest in more expensive (but more reliable) fiber-optic connections to the Internet.

The typical WISP broadcasts its Internet from towers, but some locations do not necessarily require towers. In my area if there is an ideal high location on a hill that can see several homes in an area that does not have broadband, I will collaborate with folks in the neighborhood and try to convince the person who owns the hill or high part on their home to allow me to broadcast my Internet signals to the homes in the area, and offer an exchange of free Internet to pay the rent.

Microwave Internet, most times, requires some form of line of sight from the tower to the location for delivery of the Internet signal. I say most times because I have clients who do not necessarily have a clear line of sight to the tower but receive great service. Many factors make the delivery of microwave Internet possible or not possible—for example, the distance from the tower to your site is a factor, also whether there are any obstructions to that tower, such as a hill, leafy trees, etc.

Being a relatively new industry, when you approach a WISP for service, you are probably dealing with a small business, an entrepreneur whose desire is to deliver you the very best service they can. We are a flexible group because most of us have not been gobbled up by larger corporations. This flexibility allows us to craft packages that can accommodate almost anyone's budget. WISPs are not necessarily regulated by the government. At my shop we offer monthly packages to our clients that range from $25 per month to $89 per month depending on what the speed needs are of the client. Installations require an additional one-time cost. Because these are special systems, most WISPs do not allow for self-installations. To provide a successful installation, you have to know where the towers are located and how to perform installation processes that lead to ensuring quality connections. There is no quality control on a self-installation.

The problems that can cause frustration from a service of this nature are as follows.

Trees grow in the way of the line of sight to the towers. This form of Internet is microwave, and the moisture of leaf-bearing trees can absorb the microwave signal, and your reception may deteriorate. In cases like this it is not uncommon for people to have good Internet in the winter and early spring months; then when the trees leaf up, their Internet service degrades.

Additionally, lightning can strike towers and the equipment located at your place. Wind can blow the equipment off the house, off the tower, or out of alignment. Radio interference from the wireless router in your home may interfere with your providers' ability to deliver perfect service. Aggregate this problem with the wireless routers in your neighborhood and competition from the area's wireless devices. All these factors can make the wireless provider's life interesting, to say the least.

Some folks ask if rain or snow can impact the performance. Usually, I will say no. In my area, a rainstorm will typically blow in and leave as quickly as it came. In other parts of the country where you have longer-lasting rainstorms, the rain may impact Internet performance because, like with the moisture in the leaf-bearing trees, the rain may absorb some of the microwave energy and thus impact your service. Same goes with snow.

Most WISPs only provide Internet to the LAN port on the PoE (power over Ethernet device) and you will be expected to provide and program your own router for multiple computer access, if that is what you want. In my business, I realize that most people have multiple devices that they want Internet access to, so I provide for them a router at a nominal charge and program it for secure communications. Some smaller independent WISPs will do this too; you just have to ask them. I would rather they ask me to provide one or (if they'd rather) have one of their own available on the day of the installation so that I can program it for secure communications. The reason I do this is twofold. I do not want anyone stealing their Internet connection and making their service slow, and I do not want people getting onto the client's local area network and viewing the data on their computers. The last thing I want is a client with an open network that anyone can connect to. It is bad business.

Customer service is usually local. This is very helpful to the customer. Most WISPs are generally operated by people who live in the area they work. You might have to leave a message at odd times of the day or evening, but they will usually respond and will speak regular English and are willing to help you through most problems that may not even be related to your Internet problem, such as local area networking problems or even computer-related issues.

CHAPTER 6
Satellite Internet Service

Medium: Air and space.

Pros: Availability. This type of service can be found almost anywhere in the US.

Cons: Speeds and distance for packets to travel. Due to capacity issues, this service works for those who just want fast email and fast-loading web pages. Performing any streaming for any duration (sustained downloading) will cause to you violate the fair usage policy.

Cost: Dial-up $25 to $100 per month depending on what speeds you buy and the company you go with.

Speed expectations: 1 to 12 Mbps download depending on what you are willing to pay.

In the mid to late 1990s, satellite television companies began offering Internet packets to their clients. Satellite television is already a digital transmission which means the way that the picture and sound gets to your home is already in packet form but could only be accessed through a television set. The hurdle that they faced was that Internet is a two-way form of communication, send and receive (upload and download). If you had satellite television back in the '90s, you had to plug your receiver into the telephone outlet so that the receiver could call out to the satellite company to inform the satellite company about the pay-per-view shows or movies you had watched so that they could bill you for that entertainment. For television alone that worked great, for Internet, not so great. In the infancy of satellite Internet service, you had a modem that accepted their download Internet, and you used the telephone system for the upload. The technology to receive satellite television is similar to wireless Internet services and cellular Internet services; the medium of transport is air and space, utilizing microwave radio transmission.

Since the technology is satellite, this means the service is beamed via the microwave radio transmission from their ground station, up to the satellite in space, then down to your location. This is a path of roughly 44,000 miles along which a packet of data must travel. With all other forms of Internet, your signals stay here on earth even if the medium is copper wire, fiber-optic wire, or fixed based/cellular microwave transmission.

The main benefit of a satellite broadband Internet connection is that it can be set up almost anywhere in the United States. All that is required is a clear line of site to the southern skies, in the United States. So, places that are commonly called "the middle of nowhere" usually have the ability to receive broadband Internet. The technology has changed from its infancy because now the satellite services can upload via the same link instead of using your telephone line to upload.

There are two downsides to this service. The first is the distance over which the packet must travel. Latency is the time it takes a packet of data to make it to its desired location. Latency with satellite connections is higher due to this 44,000-mile distance a packet must go to reach its destination. An Internet connection with high latency will show itself in a streaming video as buffering, or stop and starts, of the video. Low latency connections provide streaming content that does not chop up or stop and start. Since the distance that most packets have to travel on all other forms of Internet is substantially less, latency should be much less.

The other downside of this service is that since it can be found most anywhere a southern exposure to the sky is found, it is popular. Often it is the only broadband option in remote locations. Some small towns where the telephone company, the cable company, or WISPs chose not to go, you will find this form of Internet. By being popular, it is heavily used. Being heavily used, the satellite companies either need to put up more satellites to improve their capacity or limit access to people who use bandwidth-intensive applications like YouTube or Netflix. When everyone on a network is trying to watch a movie or video, everyone's performance degrades. This is true of all forms of Internet access.

Instead of expensing several hundreds of millions to billions of dollars to send up more satellites for capacity, the fix to this problem is the satellite companies enforce a policy that gives all of its users a fair share of bandwidth. This policy is called their fair usage policy. A typical fair usage policy is that if you download too much data in a given time frame, the satellite company will throttle down your connection speed because you have downloaded more than your fair share of data over that time frame. Usually that data ticker resets itself at the end of a 24-hour cycle when the infraction of downloading too much data had happened.

I hear all the time about this problem. These companies do not necessarily inform their new client about this, although it is found in the fine print of their contracts. (To see an example of a fair [or acceptable] usage policy, please see appendix D.) People find out about the fair usage policy when they are watching a movie on Netflix, and the movie just stops or buffers for an hour. Unfortunately, no one bothers to read their contracts, or if they do, maybe the fair usage policy is not clear enough about what it does, or no one bothers to ask what the gobbledygook means. The people who meet up with the fair usage policy feel trapped because they are tied to a multiyear contract.

This service is ideal for those who just want fast email, have their online banking work, and choose to live in the middle of nowhere. If your entertainment needs must be met by your broadband connection, this is not a service for you, and probably, such a remote location is not the place to move to.

The issues that satellite Internet providers usually have that cause frustration with their customers are often weather related and out of the provider's control. These kinds of problems include snow buildup on the dish that degrades (or eliminates) the signal received by the receiver at your location. Weather can also damage equipment at your location and cause service to degrade. High winds can move the equipment. Lightning can do bad things to the equipment. You might have a clear day at your location, but if the place where the origination of the Internet is experiencing difficult weather conditions, your Internet service may suffer too.

Installation of this type of Internet is typically performed by a contractor of the satellite Internet company or the company's authorized installer. Self-installation of this type of Internet is not usually in your best interest or the best interest of the satellite Internet company because precise location of the dish and receiver on your home, as well as precision pointing of the dish into space, is necessary. The installers have special equipment that they use to point the dish to the proper satellite for the best results.

The cost for this service when they began offering its service in the mid-'90s was roughly $60 per month for a standard 1.5 Mbps download. There were no tiers of speed/price back then. Now they offer different speeds at different price plans. I recently have seen their marketing ploy that they offer 4G speed. 4G speed means really nothing. 4G is a term used in cellular telephone Internet connection, more on that later. I have seen them offer 12 Mbps download for $89 per month. What I have seen in the field is an initial burst of 12 Mbps download, but the sustained download speed drops to 1 Mbps or less.

The modems that satellite companies sell are for one computer only, so if you have more than one computer, you will also need a router to distribute the Internet to those other devices. Satellite companies do not provide or support any kind of router installation, and their demarcation point of responsibility is the LAN port of the modem.

CHAPTER 7
Cellular Internet Service

Cellular Tower

Medium: Air.

Pros: Availability. This type of service can be found almost anywhere there is cellular telephone service. Speeds in urban areas can be fast.

Cons: Speeds in rural areas can be slow. Due to the popularity of this service and the cellular industry's capacity for the Internet, cellular Internet can become expensive with heavy data usage. This is often not readily recognized by the customer because the fees are clouded in legalese written in the contract that many will not readily understand.

Cost: $50 to $100 per month per phone depending on what services you choose to include in your cellular telephone package.

Speed expectations: 0.5 to 40 Mbps download depending on what you are willing to pay and the towers you connect to.

The newest entry into the Internet service provider industry is the cellular telephone companies. When cellular telephone service began to become more widely used in the late 1980s to early 1990s, the service was for voice communication and used analog technology. Analog technology is what some AM or FM radio technology use today. Back then, analog service worked well for voice communications, and still does, so some AM and FM radio technology is still in use. Others have switched to digital transmission technology. Service in those days was limited to mostly urban areas where there were large populations and thus customers. Now you can find cellular towers almost anywhere in the United States.

The technology in this form of Internet service has come a long way. At the beginning of cellular telephone communications just a voice telephone call was what was required, so the first generation of cellular communications was analog microwave transmission. This worked great. In the mid- to late 1990s, older "pager technology" married PDA (personal digital assistant) technology, and texting was born. I can remember owning a device that was a texting machine with some functionality of a PDA.

The cell phone companies and the manufacturers of the cell phones we use today took this idea further and created a phone that not only allows us to speak to someone but also to send them a text message. The next evolution, or generation (as it is called in the industry), of cellular technology was integrating the Internet into the phones bought and the services provided. As of the date of this book we are currently in the 4G-LTE evolution of cell phone technology. What does *4G* mean? What does *LTE* mean?

4G means fourth generation of a standard developed by cellular carriers and cellular telephone manufacturers. Some other providers of

Internet like to use this as a marketing ploy to get you to buy in on the hype and thus their service. I have seen the satellite Internet providers with their marketing literature that comes in the mail as stating "4G" speeds. This really means nothing, other than marketing hype. They are promoting a word that they believe everyone knows, but do not really use the true meaning of 4G. 4G is not a measure of speed. Speed is simply how fast that information gets to us through the Internet and is what the concept of "broadband Internet" is all about.

LTE means long-term evolution. It is another standard that cellular service providers and cellular manufacturers are striving to implement for today's bandwidth-intensive applications that are now going mobile. So 4G-LTE is simply stating where in the evolution of cellular technology the industry is setting its standards.

The providers of most cellular telephone Internet are large corporations. There are some smaller "large corporations" now selling service using the large corporations' capital assets (meaning their towers and electronic equipment). You will be hard pressed to find a mom-and-pop cellular company that provides cellular telephone service, much less cellular Internet service.

Items that can cause frustration when using this form of fast Internet include:

1. No cellular coverage in some sparsely populated areas.

2. The cell phone companies do not like to expand capital where there are not many paying customers, similar to the large telephone companies.

3. There are punitive costs based on data usage limits imposed for going over the allocated data limits (as defined in the small print of a contract), or when the cellular service provider company claims no data over-usage charges, they will throttle you to a slower speed, similar to what the satellite Internet companies will do to enforce their fair usage policies.

4. There are additional fees to include a hotspot access system for your other wireless toys (like a laptop computer) to gain access to the Internet. The phones required to provide this hotspot service are of the "smart phone" family of phone devices.

This service is ideal for folks who just periodically check email, their bank's websites, or Facebook. This service is not great for those who wish to view their favorite shows on Hulu or Netflix, because these streaming video sites are data intensive, meaning they download a lot of information. If you chose to use your cell phone for this type of entertainment, beware that your cell phone bill will be high.

The cost for this type of service depends on what you can shop for. This form of Internet is highly competitive. I pay $170 per month for two smart phones with unlimited texting and phone calls and 4 gigabytes of data usage. I do not go over my data plan, so I do not know exactly what the overage costs are. Remember, you must have a smart phone, and the cost of that equipment can run from $50 to several hundreds of dollars.

CHAPTER 8
Fiber-Optic Internet Service

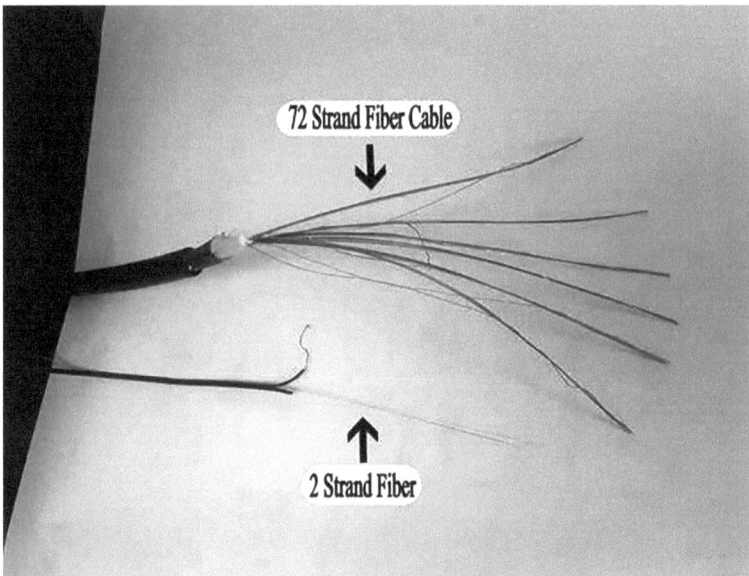

Fiber-optic Internet

Medium: Fiberglass.

Pros: Highest capacity of all services and thus fastest data speeds.

Cons: Availability, fiberglass-optic cable is not everywhere.

Cost: $25 to $100 per month depending on what speeds you buy and the company you go with.

Speed expectations: 1 to 1,000 Mbps download depending on what you are willing to pay.

The history of the concept of transmission of data using this medium goes back a few hundred years. The idea behind this form of transmission is turning on and off a light at high speeds. A simple analogy to understanding this technology is to get a garden hose, straighten it out, put a flashlight on one end, and turn it on and off while another person on the other end of the hose looks for the flashing of the light. If the light really could travel along the hose and you knew Morse code, you could interpret the communications.

The more modern medium, of course, is not a garden hose but a piece of fiberglass. It is the same type of fiberglass you use to insulate your home. The difference is that the glass is extruded into a thin wire. This extrusion is done similarly to how wool is made from sheep's hair, although in reverse. While a bundle of shaved hair from the sheep is pulled and stretched into a yarn, the glass, while in liquid form, is pushed through a tiny, round die cast. Since the extruded glass is clear, or transparent, it is then covered with a plastic material called cladding that prevents the light from escaping from the fiber.

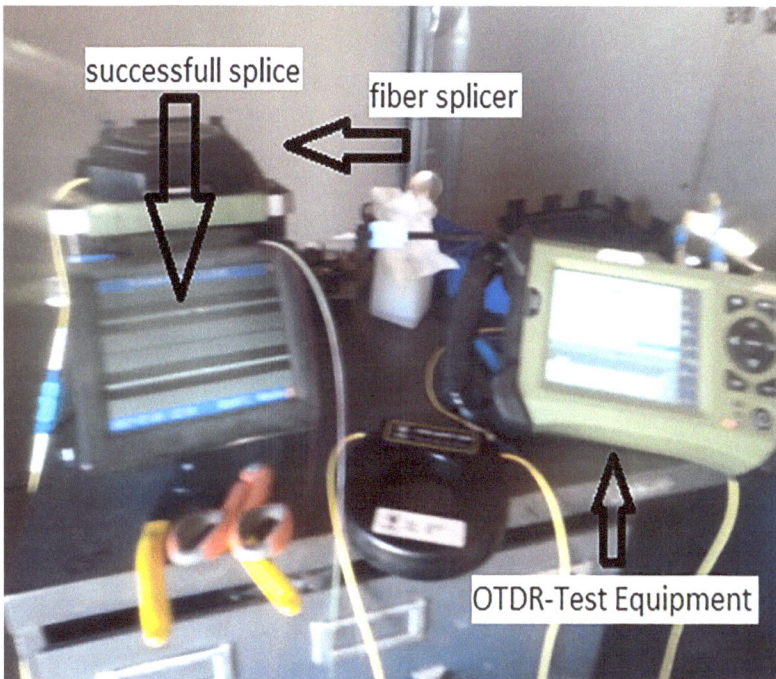

successfull splice
fiber splicer
OTDR-Test Equipment

Fiber-optic cable is about the size of a human hair. Can you imagine having to work with a human hair? In our industry, we use magnifying scopes to clean, polish, and splice these extremely small cables.

To return to our analogy, if the fiberglass cable is the hose, what is the flashlight? Today, the flashlights that send the signals are either lasers or light-emitting diodes (LEDs). Electronics turn these light sources on and off at incredibly high rates, faster than your eye can see. (Speaking of your eye, a word of warning, never ever look into a fiber-optic cable that is powered by a laser, as eye damage will occur.)

Electronics on the other end read the flashing lights and convert that flashing into either another flashing light or into an electrical signal. There are distance limitations on what the receiving end can see in the flashing before the light becomes too weak to be seen by the receiver. That is why the signal might be recreated into another flashing of light. This is how great distances can be covered.

The flashing light is regenerated every 15 to 20 miles. Where it is converted to an electrical signal is typically when it is turned into another form of Internet. In some areas where telephone companies provide DSL Internet, they will turn the fiber-optic connection from light to electric at the DSLAM to provide their DSL product. The cable companies do this conversion too. They change their Internet from fiber optics to electrical signals that go through their coaxial cable. Cellular and wireless Internet companies in many areas receive their Internet via fiber optics and convert it to electrical signals that then convert to their microwave transmissions. In some areas of the country, you may see companies running fiber optics to the home, and the conversion to an electrical signal happens at your home. You may have heard of Fios, which is a trademarked service from Verizon. This is fiber-optic Internet connections to your home.

You may be asking, why does the majority, if not all, of the providers of fast Internet use fiber optics in one way or another? The answer is that it is a reliable form of high-capacity, fast Internet. All providers of Internet buy massive amounts of fast Internet called bandwidth and break it down for consumers. The best method of purchasing these massive amounts comes across the fiber-optic medium.

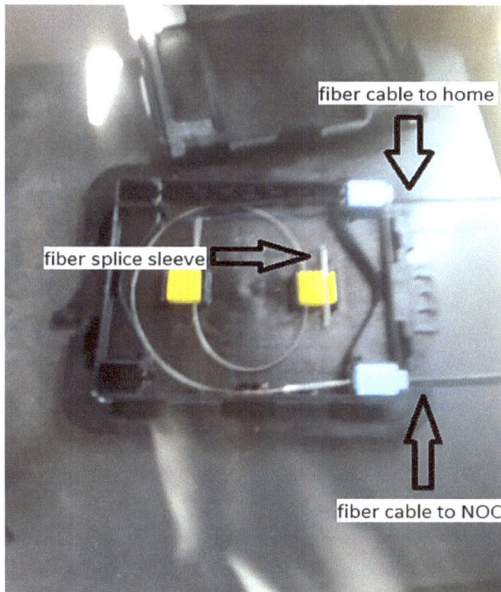

The next question you may ask is, why does fiber optics provide the most bandwidth across its medium? Most of you have seen the colors in a rainbow. Each color—red, yellow, blue, green, etc.—is a different kind of light. The advantage of fiber-optic transmission of Internet signal is that the lasers or diodes that flash the light can send the signals over the fiber-optic cable using the full spectrum of light, or each of the many colors. One strand of fiber can carry many services of Internet for many people, or they can be combined to create more capacity than when using just one kind of light.

Providers of fiber-optic service to your home are typically the telephone companies. Other providers of this form of Internet are companies who tie the whole Internet together, and their customers are typically companies that provide you your Internet (cable, cellular telephone, wireless Internet, and the head ends of satellite companies).

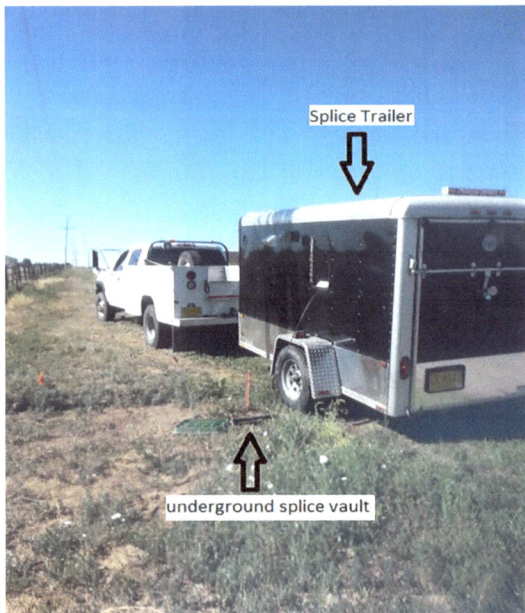

When the services are sold to other companies, the costs are high. I paid roughly $4,500 per month for a fiber-optic connection to the Internet. If you are in the area of a company that resells fiber optics to your home, you could expect to pay $50 to $100 per month.

Problems on a fiber-optic connection can include contractors or individuals digging up unseen cables and cutting them. Trees or tree limbs at times fall on the power poles and lines that the fiber-optic cable may be attached to. Lightning can damage electronics that flash the light or receive the light.

CHAPTER 9
Six Tips to Make Your Connection Better

Tip 1: Ask about Capacity

When shopping for your fast Internet, ask the provider this question: "What is the capacity of the Internet pipe providing service to your node?" (A node would be a DSLAM if you are speaking to the telephone company, or it could be the fixed-based wireless provider's upstream capacity.) The upstream capacities will surely correlate to the quality of service you receive.

What will this mean with a DSL service?

Basically, you are trying to find out whether the Internet "pipe," which provides the Internet to the DSLAM, is fiber optics or copper wire. If the pipe is fiber optics, the capacity is higher, and you will receive a higher quality of service for the reasons mentioned in chapter 8. You can determine if the node is fed by fiber-optic cable by the amount of fast Internet they are willing to sell you. If you ask what speeds you can receive and they say 1.5 Mbps / 768 Kbps, it is highly likely that you are near a DSLAM that is fed by a bunch of copper T-1 circuits. If they tell you more than 1.5 M / 768 K, then that DSLAM is likely being fed by fiber optics. You might also be able to tell if a DSLAM is fed with fiber optics by orange markers going along the roadside. These markers tell people who dig in the right-of-way that there is a fiber-optic line buried below the ground and to be careful when digging.

What this means when asking a WISP: Do they have enough capacity to meet the demand? Again, this is a question about what transport services (the "pipe") they are buying to provide Internet access into your home. Is their Internet feed a fiber-optic service or a high-capacity

FCC-licensed wireless link? If they say DSL or T-1s bonded together to provide the Internet connection for you, look elsewhere. They may be able to deliver a connection, but with the advent of bandwidth-intensive applications like Netflix, you will see performance problems in the evenings when everyone is watching Netflix.

Most cable providers' upstream services are all fiber-optically fed, so speeds should be adequate with additional growth. I say most, because most cable companies are no longer independent and are a part of the large corporate cable industry.

The satellite providers, and there are only a few, are all fed by high-capacity fiber optics at their head end. To be involved in this form of Internet delivery costs a lot of money, so these providers would not scrimp on their upstream connections.

The cellular companies will not be able to answer these questions because they have so many towers in so many places, plus your cellular service moves with you. The cell companies deliver their services like the fixed-based wireless providers do to provide remote areas with Internet, such as through a high-capacity wireless link. In places where the towers are close to the telephone company's fiber-optic networks, the cellular telephone companies hook into the fiber-optic infrastructure. Cell phone companies have been known to use the telephone company's T-1 data circuits to deliver their form of Internet as well when not near fiber-optic infrastructure but close to copper wire infrastructure.

Tip 2: Computer Upkeep

These are the top two things I see slowly bog down any fast Internet connection. Of course, there are more, but these two are quite common, and you can take care of them yourself:

1. The number of bandwidth-consuming devices on your network. If you have one computer online, it receives the full strength of the broadband. If you add your smart phones, pads, laptops, televisions, etc., your connection will slow down.

2. A computer or computers on your network that are laden with viruses or malware. Malware and viruses are basically the same. Malware and viruses are harmful pieces of software you unknowingly or knowingly may download and install on your computers.

What I mean by downloading and installing unwanted software is the following.

Nothing in life is free. Manufacturers who give software away usually partner with other companies and allow them to piggyback their software onto theirs. You download the software you want and click standard installation. This opens Pandora's box because this will install what you want with all the piggybacked stuff you do not want or need. Windows computers seem to be the most susceptible to these bugs. Not so much with computers running a Mac, UNIX, or Linux operating system, but this is not the rule. There are exceptions.

What is happening is the malware and viruses on your computer run in the background of your computer and you do not even know it. When these malicious programs are running, they are consuming bandwidth and thus making your broadband connection seem slow.

When I notice people on my network with bogged-down connection speeds, I suggest they download on their Windows computer a program called Malwarebytes Anti-Malware. Google it and use it. Most people have some antivirus programs on their machine, which is great. Use them frequently and keep the updates current. If you use Malwarebytes and your antivirus program frequently, your speed issues may slowly go away.

The big companies do not address this issue because, as I say, they are big. They will try to solve the problem by throwing more bandwidth at speed complaint issues. I have seen partnerships between the local telephone company and an antivirus company to try to alleviate this problem, but even with this partnership there are still problems because no one runs what the telephone company gives, or no one is willing to pay for the antivirus software definitions and update them like they should once the free year of antivirus subscription ends. The providers who charge by the data usage have no interest in helping you because if these problems are running on your machines, you are using data and they can charge you more money.

Tip 3: Don't Install a Server on a Residential Network

Servers to the Internet should reside in dedicated server rooms where the bandwidth is plentiful. When I first got my first taste of broadband (years before getting professionally involved in this industry), I put a server on the Internet to have fun. I did not know what I was doing, and I did not know what sort of files I wanted to serve up to the Internet (web pages, etc). It ended up being a waste of my time because the technology was over my head. Now that I know what I am doing, I look back at my behavior and must apologize to that first provider for trying to put a server on their network. Why? Because a popular server hogs bandwidth, and then everyone hates you!

Say you buy 15 to 20 megabytes of bandwidth (adequate to run a server) from the cable company and set one up. If your server becomes popular and everyone wants the content you have, your neighbors will hate you because all the people accessing your content will hog the bandwidth in the network. Your provider will hate you for the same reason. Your provider wants to provide fast Internet to all its customers. When you are gobbling up all the bandwidth because of the popularity of the content you are offering on their network, you are making everyone's service slow. To solve this problem, your provider may call to warn you, throttle your bandwidth, or shut down your service.

ISPs can tell when you have a server on your network. The larger companies may not really care, because they can just throw more bandwidth at the problem, or they may simply ignore speed complaints because of servers on their network. I explicitly explain to my clients that there are to be no servers on my network other than at my server room. Putting a server on my network and operating it as such will get you thrown off my network. I will, however, warn folks before I throw them off.

Tip 4: Avoid Peer-to-Peer Activity

Recently I saw a documentary about Napster. Napster created a way for computers networked together via the Internet to share music freely. Napster was shut down due to copyright infringement. Granted, they were not directly involved with infringement, but they provided a tool for that to happen.

What is peer-to-peer activity? Peer-to-peer activity is where you put a file into a client program and that file is then shared with anyone else who wants that file and who has the same client application. For the purposes of this discussion, a file is anything inside your computer: software, photos, music, video, etc.

Napster got into trouble because people were putting music from their CDs into their computers. Then they were putting the music content into the Napster application, and anyone who was looking for that song within the Napster network could get it for free. The tool is not the bad thing, the application of the tool was. It is similar to appropriate use of a hammer. It was designed to hit nails, not your friend.

Why should you care? When you set up peer-to-peer clients on your machine, you turn your computer into a server on the Internet, a bad thing according to tip 2. Anyone on the Internet can go to your machine and get the music, software, videos, etc. This activity floods your connections and slows your and your neighbors' Internet because any number of people are trying to get to the files on your machine.

When Napster created this, there were others who did the same. If you see eDonkey, LimeWire, FrostWire, BitTorrent (to name just a few peer-to-peer clients on your computer), you have a peer-to-peer client on your machine.

Copyright infringement is illegal. Fundamentally, it is considered theft. This is why Napster was shut down, but that does not always stop the computer user from "sharing" files in this way. Children are notorious for this activity on their own computers thinking no one will know. However, Your ISP knows because it uses peer-to-peer monitoring technology.

There are legitimate uses of peer-to-peer computer connections. Gaming devices, like the Xbox for instance, use peer-to-peer sharing for sending updates to the gaming console. On the network I operate, we have sophisticated firewalling that senses this activity and shuts it down. It is cheaper for the small business ISP to do this rather than throw more bandwidth at the problem, because bandwidth is not necessarily cheap for resellers.

Tip 5: Consider What ISP's Technicians Are Doing on a Daily Basis

The techs in small business ISPs are not just sitting around collecting all the money you send. We are constantly monitoring the entire network that delivers you fast Internet for problems that may arise or have in fact arrived. We closely monitor your connection when you report things are slow and look for activity from your network like the things I mentioned before (peer-to-peer activity, malware, viruses, etc.).

We keep on top of the regular software updates that the manufacturers of the equipment we use make to improve performance. The manufacturers send us working products, but they find glitches in the equipment when we notify them, and they make software patches to fix those glitches. We then download the software and push those updates to the equipment to ensure the problems do not continue, or if we have yet not seen them, to make sure they do not become a problem.

All ISPs suffer the effects of acts of God. We all are at the mercy of lightning, heavy rain, wind, etc. There are things we all do to mitigate the effects of lightning, but direct strikes will fry equipment. Close strikes of lightning can even fry the sensitive electronic equipment. When this happens, we all scramble to repair what has been damaged. Some instances require climbing high towers and replacing damaged equipment; others just require driving to the sites and rebooting the equipment to get it to come back up.

The other acts of God we face are when equipment gets submerged in water. We do our best in trying to prevent that from happening, but when a cable that has been in the ground for several years gets wet, Internet performance can suffer. When electronics in the field get wet from a constant barrage of rain, they too will get fried. We do our best to fix that equipment as fast as we can because we all recognize the value of the product we deliver.

It is interesting when the power goes out, such as after a lightning storm, that people are willing to be without power for a while without much complaint. When the Internet goes down though, for whatever reason, people go ballistic. All ISPs in one way or another are tied to a power grid that is susceptible to problems induced by an imperfect power system. That fact leads to tip 6.

Tip 6: Purchase and Install a UPS (Uninterruptable Power Supply)

A UPS is a device that is used to keep the power running in the event of a power outage. A UPS for your home electronics can cost from $35 to hundreds of dollars depending on what you need or want and is well worth the investment. All responsible IPSs have or use battery backups and/or generators to maintain the uptime of their infrastructure in the event of a power event. A UPS is able to do this because it is a device with a battery inside that will continue to produce power for a brief period of time when the power company cannot.

A UPS is, in some ways, similar to a surge protector or extension cord. They are good at isolating your equipment from the power grid's problems, much like a surge suppressor. A UPS works in this way: it is plugged into a power outlet in your home and brings power from your electrical service to the device. The UPS is converting that incoming power to DC to charge its internal battery, then inverts the power back to AC for your computers, modems, etc. When your electricity goes out or you unplug the UPS from the wall, it will continue to provide power to the devices plugged into them for a period of time. Since they operate on a battery and batteries can hold only a finite charge, uptime is not unlimited. Eventually the battery will be drained and will no longer provide power in the event of an extended outage.

CHAPTER 10
What Is Available in Your Area?

Even with all the knowledge you have gained from reading this guide, it would not be complete if I did not tell you how to find what is available in your area. With the proliferation of broadband in the United States, you are not left to a singular choice for your connectivity. The list below is in order of the most accessible Internet to the least.

Satellite companies reach almost 100% of the country with their form of service. Common companies that offer this service include Exede, Dish, WildBlue, and Hughes.

Almost every area in the country also has cellular service from some provider. Reach out to them to learn their coverage. The largest, most common providers are AT&T, Verizon, Sprint, T-Mobile.

Almost every area in the country has a telephone company. They may be large Baby Bell companies to small rural carriers. Reach out to them to see what is available.

Fixed wireless broadband companies are almost everywhere in the country as well. They may be a little harder to find because they are small businesses with limited marketing dollars to get their names out unlike the larger providers of broadband. Go to the trade organization www.wispa.org to find a local WISP in your area.

Cable companies are out there. Some are independently run, and some are large conglomerations. The large corporations that offer these services include Comcast and Cox Cable, to name only a couple.

APPENDIX A
The Harm of Peer-to-Peer Networking

What you do not know your children are doing could cost you millions.

What is peer-to-peer (P2P)? According to Google, it is "denoting or relating to computer networks in which each computer can act as a server for the others allowing shared access to files and peripherals without the need for a central server."

Now what does this mean in plain English? When you purchase a broadband connection, you are purchasing a connection for only your use only. When you set yourself up as a server (see again tip 3), you are providing use of your computer to others who do not pay for the connection.

P2P is like a hammer. A hammer makes a great device for hammering in nails and such, but it can be used for bad purposes. In the same way, P2P is useful to software manufacturers who develop large data programs or program updates. This is like the hammer and the nails. They want you to use P2P to obtain these large files using a P2P client because it takes the load off of one server to provide large data files by distributing them to several servers they own. Gaming applications love for you to use their P2P clients because their software patches are extremely large. Even Microsoft, with the release of their Windows 10 product, has a built-in P2P client so that you receive your updates from their various servers around the globe, as opposed to downloading the files from one location and overloading that one connection.

The part of P2P that is like the nefarious purpose of a hammer (as a weapon) is in downloading illegal music, software, or video content. This is called copyright infringement or piracy. The people who create this content copyright their hard work. About 10 years ago, Napster created this system and was smacked down by the FTC. Yes, they created a useful tool, but the illegal use of that tool lead, us to today where we, the ISPs, receive several notices from copyright lawyers telling us that someone on our network is pirating something.

Here is an example of a letter received by us because someone on our network was found to be downloading copyrighted content. (Particulars have been removed to protect those involved.)

-----BEGIN PGP SIGNED MESSAGE-----Hash: SHA1

Dear arin-admin@yourinternetprovider.net

We are writing this message on behalf of

[Large Corporate Entertainment Company].

We have received information that an individual has utilized the below-referenced IP address at the noted date and time to offer downloads of copyrighted material.

The title in question is: [Popular Movie]

The distribution of unauthorized copies of copyrighted television programs constitutes copyright infringement under the Copyright Act, Title 17 United States Code Section 106(3). This conduct may also violate the laws of other countries, international law, and/or treaty obligations. Since you own this IP address, we request that you immediately do the following:

1. Contact the subscriber who has engaged in the conduct described above and take steps to prevent the subscriber from further downloading or uploading [Large Corporate Entertainment Company] content without authorization; and

2. Take appropriate action against the account holder under your Abuse Policy/Terms of Service Agreement.

On behalf of [Large Corporate Entertainment Company], owner of the exclusive rights in the copyrighted material at issue in this notice, we hereby state that we have a good faith belief that use of the material in the manner complained of is not authorized by [Large Corporate Entertainment Company], its respective agents, or the law.

Also, we hereby state, under penalty of per-jury, that we are authorized to act on behalf of the owner of the exclusive rights being infringed as set forth in this notification.

We appreciate your assistance and thank you for your cooperation in this matter. Your prompt response is requested.

Any further enquiries can be directed to [copyright@copyrightpeople.com].

Please include this message with your enquiry to ensure a swift response.

Respectfully,

[Mr. Important, CEO
Large Corporation, Inc.
Email: copyright@largecorporation.com
Address: Somewhere Important,
Los Angeles, United States]

You may ask, why should I care? The reason ISPs care is not necessarily that an illegal act is being performed using our services. What we do care about is your connection to the Internet and how you feel about us. You judge the electric company based on its ability to provide power. If the light comes on, the power company is doing its job. You judge the water company on its ability to deliver water. If water comes out of the tap, the water company is doing its job.

You judge your Internet provider on two criteria, connection and speed. If the connection is running, we have met our first criteria. If the speed of that connection is the speed you pay for, then we have met our second criteria. When someone on your network or in your home shares files to the Internet using a peer-to-peer client, performance (speed) goes down in your home, and it also degrades the performance to any of your neighbors who happen to be on the same network. You might experience, for instance, that the Netflix movie you want to watch will not stream. This is when your ISP is unfairly judged for not meeting those two criteria.

The why to what is happening is a science in itself, so I will try to explain it the best I can. When someone sets up a computer on a home network as a server to anyone using the Internet, they are inviting the whole world into their computer to grab small fractions of content that can be reassembled on another person's computer and create a functional end product. Let's use music as an example. When you put a song on a P2P client, people from around the globe can come to your computer and grab a few milliseconds of the song. They then go to another computer somewhere else and grab another part of the song. When all parts are collected, the program assembles all these pieces of music together and then you have the full song. Why this is harmful to networks like many ISPs? When this happens on such a large scale, it is basically multiplying this one connection to grab a part of the song by hundreds to thousands of connections. You (without knowing) end up flooding the network that provides your Internet with connections from around the globe. This "flood" impacts your performance (speed) as well as the performance of your neighbors.

In my many years of providing Internet, this has been an ongoing issue since Napster created this tool. At our ISP, we recently implemented a tool that senses when many connections are being created from your home network. We have had to do this in an effort to provide our customers the speed they pay for. In the past we have had to root out this problem by hearing about speed complaints, analyzing the client's service and the service of their neighbors, and taking action. This work was like looking for a needle in a haystack and, as such, took up a lot of our time.

We had one client who we drilled down as causing a speed problem. They had two iPhones and two iPads. After visiting their home and adjusting their service, we thought we had taken care of the issue. When we returned from that visit, the problem was still happening. I then had the client shut off all devices and turn each one on to try to isolate the issue. They helped us by doing this. One at a time we brought each device back online, to see the problem was not there. I then had them browse to a few sites they regularly visited, and, bam, we found it. It was a site for a newspaper in the Midwest that had so much video content on it running all the time that when they went to this website, it would bring their connection to its knees in regard to speed and the connections of their neighbors.

They did not do this maliciously. They just did not know, and neither did we until we searched out the problem. The resolution to this problem was for them to read the content of the website they wanted and close the website when they were done. Websites like Facebook have this issue too, streaming content running all the time. People with cell phones who have data caps and use Facebook all the time were experiencing large phone bills because of this problem.

If the tool we created senses from a clients' service a "flooding of connections," it redirects the client to a page that informs them of such and that there may be a problem on their home network. This stops the flood and informs the client that they need to stop the connection with the P2P.

A typical browsing session at www.google.com creates about five to ten connections.

The Google logo and search bar are each a connection, but not a big deal. With several computers online at the same time any ISP should expect around 80 connections as well. But when a client opens their browser and they have it set to open five to ten tabs at the same time, we can expect 40 to 50 connections.

Netflix movie watching is basically a one- to five-connection load, also not a problem. When our monitoring tools senses more than 500 connections from our client's home network to the Internet over a period of time, that triggers a flag. If after a few minutes the connections do not come down, their network will be redirected to the web page that tells them there is something not right with their usage.

Clients tell me all the time, "I do not do peer-to-peer." That is mostly true. I can agree; most do not even know what this is. Heck, I barely know myself, but I know what it does to a network.

In our years of experience, we have learned that the people who generally end up on P2P sites are children. It is often just the nature of the things they like to do with their computers. Most all children now have computers. Do you, as a parent, know what is running on your child's computer?

Knowingly running a peer-to-peer client program on the network we operate is a violation of our acceptable use policy and our terms of service as seen on our website, and a cause for termination of your Internet connection.

However, we always try our best to mitigate these issues, especially if children are doing this activity. We do not want to ban you from our network when the problem is an innocent one. We understand the complexities of this issue.

What also causes this flooding of connections, besides peer-to-peer, to the Internet?

1. Malware and viruses, as Trojans, running on your computer that you may not even know about

2. Legitimate programs or applications that use a form of peer-to-peer, such as some cloud applications like Microsoft Office Live

If you receive the communication from your ISP about this issue, it does not mean you are a bad person or that you are doing anything illegal, it just is informing you that you have something going on within your network that is harmful to your speed and probably your neighbors' speed as well.

All ISPs are just trying to provide you with the best Internet connection they can and to deliver to you the connection and speeds that you pay for.

What can you do to ensure you are not harming your own connection speeds or your neighbors' connection speeds?

1. Look at your computers for peer-to-peer clients and eliminate them. Search the C: drive for anything with "wire" in it. You can do this by typing in the search bar "*wire." If you find anything with "wire" in it, remove the program.

2. If you do not have any of these applications on your machines, clean your machines up. Run an anti-malware program and antivirus program.

3. If you use legitimate programs that would like to use a peer-to-peer connection—such as a gaming system, some Cloud applications, or even the new Windows 10 operating system—please set the program or device updates in these devices to receive the software updates to anything other than through a file-sharing program.

APPENDIX B
Streaming Media Content

Why won't my Netflix or Amazon Prime work on my television? We get this question all the time, and the answer is multifaceted, as there are a number of reasons why your streaming content buffers.

The ISP I operate is an Internet service provider, the only application on the Internet we support is our own email service, if you chose to use it. Netflix and Amazon Prime are applications supported by Netflix and Amazon, which happen to ride over the ISP's data network.

In California, where I live part time, we have a 50 Mbps download fiber-optic connection to the Internet. We own a smart TV. The TV is 25 feet away from the router, so we run a wireless link from our router to the television. Sometimes it takes a few minutes for the Netflix movies we watch to sync up, but when it does, it streams rather well.

This was not always the case. After being quite frustrated that even with a 50 megabit download connection, I was getting the buffering you all may experience, I looked into why. There happened to be a software or firmware update for the television. So, I downloaded the software and put it on the TV and voila! No more buffering. Even with a massive Internet connection like we have in California, the Netflix service buffered and I did not blame the Internet provider. I had to first consider my own equipment and be sure it was up-to-date and clean.

Something else to consider, when my wife and I are watching a Netflix movie or show, we do only that. We are not watching the movie on the television and a YouTube video on our cell phones.

At my home in New Mexico, I use my Netflix account occasionally, and not with a smart TV. Typically, I watch the Netflix stream on my laptop computer. I have a few computers at the house and a smart phone as well, but I turn them all off as I am one person who can only do one thing at a time. This is also important for the fastest service possible.

On the laptop I am writing this book for you, I have the ability to turn my home's Internet speeds down to dial-up speed and up to 100 megabits. I used this ability to test on my network what was the slowest speed that would still run a Netflix movie. Not to be a hog on my own network, I throttle myself to 1 Mbps download, as that is what many of my clients purchase from me. At this speed, Netflix streams to my laptop connected wirelessly without buffering at all, even at peak hours. I then throttled myself to 768 Kbps, which is slower than 1 Mbps, and the Netflix streamed fine. Now understand this is one computer online that is constantly scanned for malware on a weekly basis. When I tried to stream at slower speeds, I began to receive the dreaded buffering.

The Internet is a two-way, if not more, connection between your computer and another computer (or server) somewhere in the world. Just like your ISP provides you with your connection to the Internet, the content providers you go to most (Netflix, Amazon Prime, etc.) are connected to another provider somewhere in the world. These content providers do not have the perfect connection to the Internet and can experience the same congestion you experience at your home. The content providers will never admit to congestion issues, if you could ever get into contact with them, but they experience the same congestion all ISPs have. I can say this with confidence because when I hear about people dealing with buffering on one content provider, I have them go to another content provider and stream their content. Lo and behold, the other content provider streams just fine.

I mentioned earlier having a clean machine when I stream content. A huge part of your connection to the Internet is the hygiene of your computer(s). At my ISP we have been offering for years to those who to take us up on it free clean-up of computer bugs. These bugs we all get run programs in the background of any machine that interacts with the Internet, and as such gobbles up useable bandwidth you may use for streaming. I simply run a free program called Malwarebytes on my computers to clear them of these bandwidth goblins or thieves.

These bugs not only affect your connection to the Internet but also may affect your neighbor's connection speeds to the Internet. Many people tell me they have antivirus programs on their computers they run all the time. I thank you if you do this. I highly advocate having these programs and running and updating frequently. In our years of experience, we have found that running a good antivirus program in conjunction with Malwarebytes help ensure you are running clean machines and not impacting yours or your neighbor's speed performance.

This is a global Internet problem. People leave their telephone company's DSL connection because of poor performance. When they leave the DSL provider and join my network, we install the Malwarebytes on their computers and clean off a majority of the reason they had problems with their telephone companies' DSL connection, and thus clean off their grief.

How are you getting these bugs on your computer if all you do is Netflix, Facebook, and email? You are not special in this regard. We all get these bugs by opening emails, clicking on things in Facebook, etc. None of us are immune who are connected to the Internet.

A good analogy to using a good antivirus and anti-malware program is like you going to doctors. The antivirus programs are your general practice doctor. You visit him with common ailments. A good anti-malware program is like your specialist who focuses in on only a part of your body that might be sick. And good preventive medicine means checking in regularly with your doctor, even if nothing seems to be wrong.

My advice to help ensure you do not receive the dreaded buffering of your favorite streaming site is as follows:

1. Keep your computers clean with a good antivirus and anti-malware program.

2. If you have a smart television that does not stream to your liking, check that the firmware on the TV is up-to-date and the devices that you use to stream the content to the television (such as a Roku box) are regularly rebooted (power unplugged for 10 seconds).

3. Enjoy your favorite streaming site together as a family with all other devices shut off.

4. Stay away from companies that offer bursting. Bursting is just a fancy way of saying when you run a speed test you may burst to a higher speed, but under sustained data usage you will drop to a slower speed. This plays havoc on your streaming services because these streaming services judge your connection when you start them. When you burst a fast connection to these content providers, they think you have that burst speed all the time and they will modulate the rate they deliver the content to you at that burst speed. When your speed settles to the normal rate, that confuses the content provider, and the dreaded buffering will begin.

5. Most streaming content providers will offer different ways in which you receive their content. You may see standard definition to ultrahigh definition. Pick the standard definition, and that should eliminate the dreaded buffering.

APPENDIX C
The Demarcation Point

The demarcation point is the point up to which your ISP is responsible for your service. All ISPs and utilities have a point in the overall system to which they are responsible for delivering service. Beyond this point, the responsibility falls with the end user. In other words, with you.

The demarcation point for DSL, cable, satellite, and fiber-optic Internet services is usually a LAN port located on the modem. If the providers of these services can plug their laptops into a LAN port of the modem and receive Internet, your service works to their point of responsibility. If your computer will not connect to the Internet through a wireless LAN or will not connect directly to the modem, the company may refer you to the manufacturer of your computer or manufacturer of the wireless router for help. They all have ways of verifying that they have provided service to their demarcation point without coming to your home, and when they verify this to be true, but you still cannot use the Internet on your devices, they will not fix your problem. That answer can be very frustrating, but they are simply not responsible to fix it because of the demarcation point.

Frustrating as it is, a comparison to other utilities that bring services to your home will help explain why this is so. Similar to your power company, if there is power at the panel's main breaker, they have done their job in delivering power to your home. To fix a power problem, you call an electrician. No power because a circuit breaker keeps tripping is not the responsibility of the power company. If, however, they are delivering less than 220 V of electricity to your home, that is their responsibility, and they must fix that problem.

Also similar to a water company, if water comes out their meter, they have fulfilled their responsibility in providing water. You would then call a plumber to diagnose why you have no water.

If you are having problems with your Internet or find you have no service, you will call your ISP for assistance. If your ISP happens to be a WISP, they may ask you to plug your computer directly into the LAN port of the PoE and run a speed test. These two procedures verify connection and performance to the network (remember ISPs are judged on two parameters, connection and speed). If the test comes back indicating speed is close to what you pay for, then the problem will be, nine times out of ten, related to your router or computer.

An ISP may have provided you with a Wi-Fi router when your service was installed because the industry realizes most households have more than one device to connect to the Internet. What you do with this router is your responsibility, whether you put 1 to 100 devices behind it, change encryption, ensuring proper and current software or firmware.

The routers my ISP offers have been proven to work best on our network. We will offer free technical advice via the telephone to troubleshoot router or some computer-related issues. When we have to deploy a technician to our client's home and the technician determines the problem is with your router or computer (we plug directly into the LAN port of the PoE and run a successful speed test), service charges typically apply for the visit. This is because the router and computer are beyond our demarcation point for providing service.

The big difference between a WISP and the power or water company is that we often can help you find the way to fix your problem over the phone. If you call the power or water company because of a tripping breaker or a leaky toilet, they probably know how to fix those kinds of problems but will not assist you in fixing the problem.

APPENDIX D
Fair Use Policy (Example)

Following is the acceptable use policy I use with my customers. This is posted online and is provided to each customer when they sign up for service. Also known as a fair use policy, this text is standard information for most Internet service providers. If your ISP does not have a Fair Access Policy, Acceptable Use Policy and a Privacy Policy, I suggest not doing business with them. These are standards the industry has come up with to cover their assets and protect their users.

Acceptable Use Policy

This Acceptable Use Policy (AUP) has been established in the spirit of providing guidelines for our valued customers and clarifying their rights and responsibilities as users of our services and the Internet.

Higher-Speed Internet LLC, d/b/a Higher-Speed ("Higher-Speed," "Company," "we," "our," or "us"), is an Internet Service Provider (ISP) specializing in providing high-speed Internet access and services to rural communities in the southwest. As a responsible Company, we have certain legal, ethical, and operational obligations in order to completely fulfill your needs as our customer (hereinafter "Customer," "Subscriber," "you," or "your") for exceptional and reliable service and to meet the challenge of supporting the Internet as a diverse forum for free and open discussion and the dissemination of information.

We aim to clearly outline those activities that are harmful to the efficiency of our services and that, according to our discretion, compromise our systems and/or business. Our goal includes preventing those activities that are either harmful or disruptive to the Internet activities of others. We also proscribe all activities that fall outside the boundaries of acceptable Internet use. We reserve the right to take preventative or corrective action, at our discretion, in response to anything we deem harmful or in violation of this AUP.

The use of our services by our Customers constitutes an acceptance of the terms of this AUP and our Terms of Service. This AUP may be revised in the future in order to better meet the needs of the changing Internet environment, the legal landscape in the United States and internationally, and our Customers. Your continued use of our services constitutes an acceptance of any new terms and conditions, and we attempt to, but are not required to, inform you of changes prior to implementing them. We strongly encourage you to review this AUP periodically.

Proscribed Activities

The following activities are deemed a violation of this AUP:

Illegal activities. Engaging in activities that are determined to be illegal in the United States.

Harmful activities. Attempting to harm, unduly burden or render non-operational, the accounts, computers, websites or Internet activities of others (i.e., through mail bombing, DoS or DDoS attacks, hack attacks, etc.).

File-sharing programs. Serving of files including, but not limited to, peer-to-peer (P2P) (e.g., Kazaa, Grokster, eDonkey, Ares, BitTorrent, LimeWire, etc.), client-server (i.e., an FTP server or web host), and similar automated mechanisms.

Spamming. Sending, assisting, or commissioning the transmission of unsolicited commercial email unless it conforms to the requirements of federal law under the CAN-SPAM Act. We reserve the right to take action against spamming complaints, even if the allegations meet CAN-SPAM requirements, if the allegation threatens our ability to transmit and receive email on behalf of our Customers.

Forging email headers. The transmission of email that hides or falsifies the identity of the sender.

Fraud. Negligently, recklessly, or knowingly making false or misleading statements, including but not limited to obtaining money, credit card payments, or personal identifying information through false pretenses.

Harassment. The act or intention of intimidating, threatening, or otherwise harassing others. Harassment can result from the language used, or the frequency or size of the messages.

Prohibited communications. Transmitting defamatory, harassing, abusive, or threatening language or anything that a reasonable person would regard as hate speech or literature. This includes language or other expression that significantly prejudices, creates a hostile bias, or grossly defames a group, and applies to any speech category unprotected by freedom of speech and expression.

Facilitating a violation of this AUP. Advertising, transmitting, or otherwise making available any software, program, product, service, or information that is designed to violate or assist in the violation of this AUP.

Intellectual property violations. Engaging in any activity that infringes or misappropriates the intellectual property rights of others, including but not limited to copyrights, trademarks, service marks, trade secrets, software, and patents held by other individuals, corporations, or entities. Common instances leading to intellectual property violations involve the unauthorized use of pictures and using another's trademarks without their permission to promote competing goods or services.

Violating security. Releasing your username and/or password to third parties or in any other way violating the letter or spirit of our privacy statement.

Violating US export laws. Customers are required to comply with US export control laws regardless of where they may reside. Customers may not export items prohibited by the Department of Commerce Commodity Control List.

Other harmful activities or information. Any activity or information, whether lawful or unlawful, that we deem harmful, offensive, controversial, infamous, or other to either the Company, its customers, or third parties, such that we reasonably believe our customers, operations, reputation, goodwill, or general customer relations could potentially be negatively affected.

Remedies and Action

Responsibility of avoiding the above harmful and/or unlawful activities rests solely on our Customer and authorized account members. We do not, and will not, monitor or investigate the communications of our customers.

When we become aware of a violation of our AUP, we will respond in the manner we deem appropriate and according to our sole discretion. The type of action taken will depend in part on the legal risk and requirements, extent of the violation, as well as perceived breadth and severity of the harm to us or others.

We may take any action to stop the harmful activity, including but not limited to removing or blocking access to material, shutting down an account, blocking offending transmissions, requiring a monetary deposit as assurance against future behavior (i.e., "security deposit"), deleting the account such that all information is permanently and irretrievably removed from our servers, potentially without your knowledge or notice, or any other action we consider appropriate.

No credits or refunds will be issued for downtime incurred or services paid for in advance if the account is suspended (i.e., "deactivated") or deleted due to what we perceive is a violation of this AUP, whether or not it is later proven that an actual AUP violation occurred.

Privacy Policy

We at Higher-Speed Internet know you care about how Your personally identifiable information ("Personal Information") is used and shared, and We take Your privacy seriously. Please read the following to learn more about Our Privacy Policy.

By using Our Service (as defined in Your Terms of Service Agreement, available here: www.higherspeed.net or accessing Our website in any manner, You acknowledge that You accept the practices and policies outlined in this Privacy Policy, and You hereby consent that We will collect, use, and share Your Personal Information in the following ways.

Remember that Your use of the Service is at all times subject to Your Service Agreement, which incorporates this Privacy Policy by reference. Your use of Our website is at all times subject to Our Website Terms of Use. Any capitalized terms not defined in this Privacy Policy will have the same meaning as defined in Your Service Agreement and Website Terms of Use.

Our Services are designed and targeted to U.S. audiences and are governed by and operated in accordance with the laws of the U.S. If You are not a U.S. citizen or do not reside in the U.S., You voluntarily consent to the collection, transfer, use, disclosure and retention of Your Personal Information in the U.S. You also agree to waive any claims that may arise under Your own national laws. Our Services are designed and targeted to U.S. audiences and are governed by and operated in accordance with the laws of the U.S. If You are not a U.S. citizen or do not reside in the U.S., You voluntarily consent to the collection, transfer, use, disclosure and retention of Your Personal Information in the U.S. You also agree to waive any claims that may arise under Your own national laws.

When you use the Service, the Personal Information (as defined below) You send and receive is transmitted over a wireless network or fiber network and may be subject to interception by unauthorized third parties who seek to do you harm. While it is Our objective to take reasonable measures to reduce the risk that unauthorized third parties will be able to intercept the information you send and receive through the Service, We cannot and do not make any guarantee that transmissions over the Internet are 100% secure or error-free.

We recommend that you use caution when sending any Personal Information over the Internet and use encryption technology whenever possible, such as websites that have the "https" designation in the website's address bar and show a padlock icon in the browser's window.

We do not knowingly collect, solicit or use Personal Information from anyone under the age of 13. If You are under 13, please do not attempt to register for the Services or send any Personal Information about yourself to us. If we learn that We have collected Personal Information from a child under age 13, We will delete that information as quickly as possible to the extent technically feasible. If You believe that Your child under 13 may have provided Us Personal Information, please contact Us at info@higherspeed.net.

What types of information does this Privacy Policy cover? We collect various types of information about You and Your use of the Service via Our website, Help Desk and call centers, postal mail, remote kiosks, Our Facebook Page or other social network platforms or by other means, generally classified as Personal Information and Non-Personal Information.

Generally, We gather and use Personal Information internally in connection with providing the Service to You, including to personalize, evaluate and improve the Service and Our ability to provide the Service to You, to contact You, to respond to and fulfill Your requests regarding the Service, and to analyze how You use the Service.

We may share Your Personal Information with Our Affiliates and with other third parties as described below.

What Information do We collect and how do We use this Information?

PERSONAL INFORMATION

Personal Information is the information You provide to us voluntarily or passively through Your use of Our Service and/or website, and which is directly associated with or reasonably linked to a specific person, computer or device. For example, through the registration process, when the equipment to provide the Service is installed, maintained or upgraded at Your premises, when You contact Us regarding the Service, and through Your account settings, We collect Personal Information such as Your name, email address, phone number, billing address and billing information (such as credit card account number or other financial account information), service address, and the nature of any of Your devices or other property making use of the Service. You may be required to provide certain Personal Information to Us in order to register with Us, to assist Us in improving Your Service or troubleshooting problems You are experiencing with the Service, Your computer or device, or otherwise to improve the quality of the Service.

We will communicate with You if you've provided Us the means to do so. For example, if you've given Us Your email address or phone number, We will email or call You about Your use of the Service or product improvements or upgrades, and other transactional information about Your Service.

We may also combine Your Personal Information with additional Personal Information obtained from Our Facebook Pages or other social network platforms, Our Affiliates, Our Operational Service Providers (third party owned companies that provide or perform services on Our behalf, to help serve You better and to perform functions in order to support Our businesses and operations), or other companies, such as credit bureaus, background check firms, and marketing research companies.

Some forms of Non-Personal Information as described below will be classified as Personal Information if required by applicable law or when such information is directly associated with or reasonably linked to a specific person, computer or device, or is combined with other forms of Personal Information.

NON-PERSONAL INFORMATION

Website Information, Use of Cookies and other Similar Tracking Technology - When you visit Our website, We will collect various types of Non-Personal Information, such as information on Our server logs from Your browser or device, which may include Your IP address, unique device identifier, "cookie" information, the type of browser and/or device you're using to access the Service, and the page or feature You requested. (IP Address and device identifiers are traditionally classified as Non-Personal Information, unless We are required to do so otherwise under applicable law.) "Cookies" and "web beacons" are text file identifiers We transfer to Your browser or device that allow Us to recognize Your browser or device and tell us how and when pages and features on Our website are visited, by how many people, and other activity on the website.

You can change the preferences on Your browser or device to prevent or limit Your device's acceptance of cookies, web beacons or other similar technology, but this may prevent You from taking advantage of some of the features on Our website, or accessing certain functions and conveniences. If You click on a link to a third-party website or service, such third party may also transmit cookies to You. Again, this Privacy Policy does not cover the use of cookies or other such tracking technology by any third parties, and We are not responsible for their privacy policies and practices.

We also use Personal Information and Non-Personal Information to enhance Our website and Our Service offerings. For example, such information can tell Us how often visitors use a particular feature of Our website and which products and services are most interesting to current and potential customers, and We can use that knowledge to make the website useful and interesting to as many users as possible and to enhance and refine Our Service offerings. We will continue to conduct analytics on Our website performance; You may not opt-out of this use of cookies or other Personal Information or Non-Personal Information.

Technology is improving every day and to improve Our Services' operation and function We may introduce new technologies and monitoring techniques without advance notice or consent from You. We may also use third party providers to conduct such internal analyses.

NETWORK INFORMATION

We also collect Network Information, information about Your access to, and use of, Our network, which may or may not be directly associated with or reasonably linked to a specific person, computer or device. For example, We may collect information about the performance of the Provider Equipment installed on Your property or at Your premises, when You are using the Service, the various devices You are using to access the Service, the amount of data You are transmitting and receiving, the content of the data You are transmitting and receiving, the websites You are visiting, and any other information that is transmitted over Our network. We may also aggregate Network Information from multiple subscribers, and We will share such aggregated Non-Personal information about the overall performance of Our Service and network with Our Affiliates and other third parties. Aggregated information does not identify a specific individual, computer of device.

How is Network Information Used?

We use Network Information to monitor and enhance the performance of Our network. We will not monitor the content of the websites viewed or email communications as part of Our standard network management. Generally, We will only monitor and preserve the following Network Information:

- When You are using the Service.

- How You are using the Service, such as monitoring traffic patterns regarding websites visited, amount of data being sent or received, or other activity.

- The amount and types of data You are transmitting and receiving through the Service; and

- General information regarding the performance of the Provider Equipment installed on Your property or at Your premises, and its interaction with the rest of Our network.

However, We reserve the right to, and may, monitor, access, review and preserve any Network Information and/or content in the following situations:

- In response to an inquiry from You or another Authorized User on Your account regarding Your or their use of the Service or problems You or they are experiencing using the Service.

- If We have reason to believe You are using the Service in violation of Your Service Agreement or any applicable statutes, rules, ordinances or regulations.

- If We have reason to believe Your use of the Service is negatively affecting other Customers; or

- When We are required by law or legal process to do so, or when We have a good faith belief that We are required by law or legal process to do so.

How is Personal Information used for marketing and advertising purposes?

We will use Personal Information to send You marketing and advertising messages related to Our Service and Our website using Your email address, postal address, or Your telephone number (for voice, texts, and pre-recorded calls). We may deliver a marketing or advertising message based on Your visits to Our website, which will be general advertising or "Contextual Advertising," which is advertising based on the subject matter or the content of the specific website page or subject matter. We may also send You "First Party Advertising," which is advertising that is customized or personalized based on a history of Your use of our Services (possibly combined with information from our Facebook fan page or other social network platforms). First Party Advertising is based solely on a combination of information We collect from You – not from Your visits to other websites across the Internet.

You may opt-out of First Party Advertising but not Contextual Advertising. No Personal Information is used to deliver Contextual Advertising; it automatically will appear based on the content or webpage you are viewing. And You may continue to receive general advertising if You opt-out of First Party Advertising, it will not be customized or personalized for You.

We do not provide third party "Network Advertising," which is advertising based on Your overall Internet usage across different third party websites or online services. Multiple third-party websites and online services are involved in this tailored or personalized advertising process, in essence a "network" of advertising providers.

Because We do not provide network ads, We do not recognize the "Do Not Track" settings on various Internet browsers. We do not engage or allow third parties to track you across the Internet and across time for advertising purposes.

Links to other websites or online services.

Our website and/or Facebook Pages (or other social networking platforms) may contain a variety of content and functionality and may provide links to other third-party websites or online services. Despite such links, this Privacy Policy applies only to Us and Our Affiliates. The presence of a link does not constitute or imply Our endorsement, recommendation, or sponsorship of the content, goods, services, business or privacy practices on such websites or online services. We encourage You to be aware and informed when You leave Our website and Our Facebook Pages, or any other social networking platforms.

Will We share Your Personal Information?

Your Personal Information will only be disclosed to third parties (including Our Affiliates) as listed in this Privacy Policy, or if We have received your consent at the time We collect your Personal Information or prior to the disclosure of any Personal Information. We reserve the right to fully use, disclose and process any Non-Personal Information collected from You in any manner as well as any information You make public via Our Services or website. We will not rent, sell, nor disclose Personal Information to anyone not related to Us for marketing or promotional purposes, unless We change ownership via a sale, merger or a corporate restructuring. (See "For Business Transfers" below for more information.) We will share Your Personal Information with Our Affiliates and with other third parties as described in this section for the following reasons:

- To Our Affiliates: We rely on various Affiliates in order to provide the Service to You. These are companies that are related to Us by common ownership or control. We may share Personal Information and Non-Personal Information with any of Our Affiliates for business, operational, promotional and/or marketing and advertising purposes.

- To Operational Service Providers: We and Our Affiliates contract with other companies and people to perform tasks or services on Our behalf and need to share Your Personal Information to provide products or services to You. For example, We may use a payment processing company to receive and process Your ACH or credit card transactions for Us, or We may contract with third parties to assist Us in optimizing Our network. Unless We tell you differently, We do not grant Our Operational Service Providers any right to use the Personal Information We share with them beyond what is necessary to assist Us.

- For Business Transfers/Restructuring: We may choose to buy or sell assets, or We may be sold. In these types of transactions, customer Personal Information is typically one of the business assets that would be disclosed and transferred. Also, if We (or Our assets) are acquired, or if We go out of business, enter bankruptcy, or go through some other change of control, including restructuring, reorganization or financing arrangements, Personal Information could be one of the assets disclosed, transferred to or acquired by a third party.

- For Our Protection, and the Protection of Our Affiliates, Employees, Operational Service Providers, Users and Subscribers and Public Safety: We reserve the right to access, read, preserve, and disclose any Personal Information We have access to if We believe doing so will implement and/ or enforce or Service Agreement, Website Terms of Use Agreement, Privacy Policy or any legal document; protect our Network(s), website(s), and company assets; protect the interests, rights, property, and/ or safety of Provider or Our Affiliates, employees and officers/ directors, Operational Service Providers, Users and Subscribers, agents, third party licensors or suppliers, or the general public.

- When Required by Law or in Response to Legal Process: We reserve the right to access, read, preserve, and disclose any Personal Information to which We have access if We are required by law or legal process to do so, or if We have a good faith belief that We are required by law or legal process to do so.

Is Your Personal Information secure?

We endeavor to protect the privacy of Your account and other Personal Information We hold in Our records using reasonable administrative, technical and physical security measures. However, We cannot and do not guarantee complete security. Unauthorized entry or use, hardware or software failure, and other factors, may compromise the security of Personal Information at any time.

Your account is protected by a password for Your privacy and security. It is Your responsibility to prevent unauthorized access to Your account and Personal Information by selecting and protecting Your password and/ or other sign-on mechanism appropriately and limiting access to Your computer, tablet or device and browser by signing off after You have finished accessing Your account. You are required to notify us immediately if Your password or account has been disclosed to a person whose name does not appear on your account, even if you have allowed such disclosure. You understand, acknowledge and agree that You are solely responsible for any use of Our Services via Your username and password.

Additionally, if You contact Us, We will ask You for verification of Your identification and account. We will not send an email or text, nor should You respond to any email or text communications asking for any sensitive or confidential Personal Information, such as social security number, bank account or credit card account number, or a driver's license number. If You receive an email or text requesting any such information from Us or someone that claims they are with Us or Our Affiliates please contact our privacy administrator immediately: info@higherspeed.net.

For Our IT Support Services as detailed in our Service Agreement, the code that allows Us to access Your computer desktop to help your resolve technical problems is limited only for that specific session. We are not able to access Your Computer without your knowledge, affirmative consent and involvement.

What Personal Information Can You access, modify and/or delete?

Generally, You may access the following Personal Information in Your account:

- Full name
- Username and password
- Email address
- Telephone number; and
- Billing and Service address
- Account and billing information

By contacting Us at info@higherspeed.net or through any online access portal We may create to enable You to view and modify Your account settings, You may access, and, in some cases, edit or delete the Personal Information listed above. For example, we may retain historic email, billing and/or Service addresses for security and verification purposes. You may not delete such information.

When You update Personal Information, however, We may maintain a copy of the unrevised information in Our records for internal security reasons and recordkeeping. Some information may remain in Our records after it is modified, amended or deleted by You or Us. We may use any aggregated data derived from or incorporating Your Personal Information after You update or delete it, but not in a manner that would identify You personally. We may also maintain Personal Information regarding You and Your use of the Service after You are no longer Our customer as required by Our business practices, by law, and/or tax reporting purposes.

The information You can view, update, and delete may also change. If You have any questions about viewing or updating information, We have on file about You, please contact Us at info@higherspeed.net.

What third party disclosure choices do You have?

You can always choose not to disclose Personal Information to Us; however, certain Personal Information is necessary for Us to provide the Service to You. You may opt-out of sharing Personal Information with Our Affiliates only for marketing or advertising purposes, but not for business or operational purposes.

You may opt-out of email marketing and advertising from Us or Our Affiliates using the "Unsubscribe" mechanism in each email. Before We send You a text for any reason or send You a pre-recorded call that contains advertising or marketing information, We will secure Your prior written express consent, which can be given via a voice recording, email, text message, postal mail, or telephone key press. Non-telemarketing pre-recorded calls do not require Your prior express consent in writing, unless they are sent to a wireless device. You understand, acknowledge and agree that such texts and pre-recorded telemarketing calls may be sent using an auto dialer and are not conditioned on your purchase of the Service. You may opt-out of receiving text messages any time by replying "STOP" or "UNSUBSCRIBE" to the text message. You may opt-out of receiving pre-recorded calls by the opt-out instructions in the call. However, You will continue to receive calls related to debt-collection and Your current Service.

You may also opt-out of First Party Advertising, but not Contextual Advertising, as detailed in the "Use of Personal Information for Marketing or Advertising Purposes" section above. You may not opt-out of Our use of cookies or other similar technology, or use of Your Personal Information and Non-Personal Information for Our internal analytics used to monitor activity on Our website, measure Our Service performance, or to operate and protect Our network.

Will this Privacy Policy ever change?

Yes, We are constantly working to improve the Service, so We will need to update this Privacy Policy from time to time as Our business practices change and service offerings increase, and/or there are changes in local, state or federal laws. Additionally, We will also make stylistic, organizational and/or grammatical changes to present Our privacy practices in a user friendly easy to read manner. We will alert You to any such changes by placing a notice on www.higherspeed. net with the effective date of the revised Privacy Policy, and/or by sending You an email, or by some other means to the extent required by law. Please note that if You haven't provided Us with Your email address or You have not updated Your contact information, those legal notices will still govern Your use of the Service, and You are still responsible for reading and understanding all notices posted on Our website. Your continued use of the Service or website after notice of any changes have been provided will indicate your acceptance of such changes, except where further steps are required by applicable law.

Use of Your Personal Information is primarily governed by the Privacy Policy in effect now that you are subscribed the Service or need visited Our website. If We elect to use or to disclose Personal Information that identifies You as an individual in a manner that is materially different than that stated in the Privacy Policy in effect at the time you subscribed to the Service or visited Our website, we will provide You with an opportunity to consent to such use or disclosure. Depending on the circumstances, that consent may include an opt-out.

What if You have questions or comments about this Privacy Policy?

If You have any questions or concerns regarding Our privacy practices and policies, please contact Us at info@higherspeed.net. © Copyright 2017 Internet Services, LLC – All Rights Reserved.

ABOUT THE AUTHOR

Steven Grabiel is a provider of Internet connections to clients in rural New Mexico. The form of Internet he provides is mostly fixed wireless and fiber-optic connections. Steven Grabiel hosts a weekly Internet radio show, www.ispradio.com, that deals with topics of interest to those in his industry.